Johann Schumann and Yan Liu (Eds.)

Applications of Neural Networks in High Assurance Systems

Studies in Computational Intelligence, Volume 268

Editor-in-Chief
Prof. Janusz Kacprzyk
Systems Research Institute
Polish Academy of Sciences
ul. Newelska 6
01-447 Warsaw
Poland
E-mail: kacprzyk@ibspan.waw.pl

Further volumes of this series can be found on our homepage: springer.com

Vol. 247. Monica Bianchini, Marco Maggini, Franco Scarselli, and Lakhmi C. Jain (Eds.)
Innovations in Neural Information Paradigms and Applications, 2009
ISBN 978-3-642-04002-3

Vol. 248. Chee Peng Lim, Lakhmi C. Jain, and Satchidananda Dehuri (Eds.)
Innovations in Swarm Intelligence, 2009
ISBN 978-3-642-04224-9

Vol. 249. Wesam Ashour Barbakh, Ying Wu, and Colin Fyfe
Non-Standard Parameter Adaptation for Exploratory Data Analysis, 2009
ISBN 978-3-642-04004-7

Vol. 250. Raymond Chiong and Sandeep Dhakal (Eds.)
Natural Intelligence for Scheduling, Planning and Packing Problems, 2009
ISBN 978-3-642-04038-2

Vol. 251. Zbigniew W. Ras and William Ribarsky (Eds.)
Advances in Information and Intelligent Systems, 2009
ISBN 978-3-642-04140-2

Vol. 252. Ngoc Thanh Nguyen and Edward Szczerbicki (Eds.)
Intelligent Systems for Knowledge Management, 2009
ISBN 978-3-642-04169-3

Vol. 253. Roger Lee and Naohiro Ishii (Eds.)
Software Engineering Research, Management and Applications 2009, 2009
ISBN 978-3-642-05440-2

Vol. 254. Kyandoghere Kyamakya, Wolfgang A. Halang, Herwig Unger, Jean Chamberlain Chedjou, Nikolai F. Rulkov, and Zhong Li (Eds.)
Recent Advances in Nonlinear Dynamics and Synchronization, 2009
ISBN 978-3-642-04226-3

Vol. 255. Catarina Silva and Bernardete Ribeiro
Inductive Inference for Large Scale Text Classification, 2009
ISBN 978-3-642-04532-5

Vol. 256. Patricia Melin, Janusz Kacprzyk, and Witold Pedrycz (Eds.)
Bio-inspired Hybrid Intelligent Systems for Image Analysis and Pattern Recognition, 2009
ISBN 978-3-642-04515-8

Vol. 257. Oscar Castillo, Witold Pedrycz, and Janusz Kacprzyk (Eds.)
Evolutionary Design of Intelligent Systems in Modeling, Simulation and Control, 2009
ISBN 978-3-642-04513-4

Vol. 258. Leonardo Franco, David A. Elizondo, and José M. Jerez (Eds.)
Constructive Neural Networks, 2009
ISBN 978-3-642-04511-0

Vol. 259. Kasthurirangan Gopalakrishnan, Halil Ceylan, and Nii O. Attoh-Okine (Eds.)
Intelligent and Soft Computing in Infrastructure Systems Engineering, 2009
ISBN 978-3-642-04585-1

Vol. 260. Edward Szczerbicki and Ngoc Thanh Nguyen (Eds.)
Smart Information and Knowledge Management, 2009
ISBN 978-3-642-04583-7

Vol. 261. Nadia Nedjah, Leandro dos Santos Coelho, and Luiza de Macedo de Mourelle (Eds.)
Multi-Objective Swarm Intelligent Systems, 2009
ISBN 978-3-642-05164-7

Vol. 262. Jacek Koronacki, Zbigniew W. Ras, Slawomir T. Wierzchon, and Janusz Kacprzyk (Eds.)
Advances in Machine Learning I, 2009
ISBN 978-3-642-05176-0

Vol. 263. Jacek Koronacki, Zbigniew W. Ras, Slawomir T. Wierzchon, and Janusz Kacprzyk (Eds.)
Advances in Machine Learning II, 2009
ISBN 978-3-642-05178-4

Vol. 264. Olivier Sigaud and Jan Peters (Eds.)
From Motor Learning to Interaction Learning in Robots, 2009
ISBN 978-3-642-05180-7

Vol. 265. Zbigniew W. Ras and Li-Shiang Tsay (Eds.)
Advances in Intelligent Information Systems, 2009
ISBN 978-3-642-05182-1

Vol. 266. Akitoshi Hanazawa, Tsutom Miki, and Keiichi Horio (Eds.)
Brain-Inspired Information Technology, 2009
ISBN 978-3-642-04024-5

Vol. 267. Ivan Zelinka, Sergej Celikovský, Hendrik Richter, and Guanrong Chen (Eds.)
Evolutionary Algorithms and Chaotic Systems, 2009
ISBN 978-3-642-10706-1

Vol. 268. Johann Schumann and Yan Liu (Eds.)
Applications of Neural Networks in High Assurance Systems, 2010
ISBN 978-3-642-10689-7

Johann Schumann and Yan Liu (Eds.)

Applications of Neural Networks in High Assurance Systems

Springer

Dr. Johann Schumann
RIACS/USRA
NASA Ames Research Center
M/S 269-3
Moffett Field, CA 94035
USA
E-mail: Johann.M.Schumann@nasa.gov

Dr. Yan Liu
Motorola Autonomic Networking Labs
1301 East Algonquin Road
Schaumburg, IL 60193
USA
E-mail: yanliu@motorola.com

ISBN 978-3-642-10689-7 e-ISBN 978-3-642-10690-3

DOI 10.1007/978-3-642-10690-3

Studies in Computational Intelligence ISSN 1860-949X

Library of Congress Control Number: 2009942253

Typeset & Cover Design: Scientific Publishing Services Pvt. Ltd., Chennai, India.

Printed in acid-free paper

9 8 7 6 5 4 3 2 1

springer.com

Preface

This book is the first book available which directly addresses a key part of neural network technology: methods used to pass the tough verification and validation (V&V) standards required in many safety-critical applications.

Years ago, Chuck Jorgensen of NASA Ames explained what was then one of the most impressive real-world uses of neural network technology: landing a real MD–11 on automatic pilot, with the hydraulic surfaces all locked up, in a way which made it impossible even for a human to land according to the judgment of many engineers at the time.

The lesser accomplishment was the landing. The greater accomplishment was to pass the tough but reasonable V&V standards needed to get permission to do the experiment. NASA Dryden, which did much of the V&V evaluation, worked hard with NASA Ames to develop new, rigorous standards for evaluation suitable for a new nonlinear technology. For others working on neural networks for safety-critical applications, it is essential to know what kinds of evaluation methods have been developed across many sectors, and how to pass the tests.

Of course, this is only one of several essential strands of research. In the university world, theorems are more popular as a method for evaluating and understanding new control systems. As this book goes to press, new stability theorems for Heuristic Dynamic Programming have been proved by Frank Lewis and Jagannathan Sarangapani, presented at the International Joint Conference on Neural Networks 2009. Suykens et al proved stability theorems years ago for simpler methods of neurocontrol, far stronger than the stability theorems available for ordinary adaptive control even in the linear case. But when systems are complex and nonlinear enough, absolute formal mathematical guarantees become impossible, no matter what the control design, if the plant and the controller are modeled in a fully accurate way, reflecting the full uncertainties. In the aircraft industry, there is a joke that people first develop a provably stable controller—and then spend 80% of the project budget tweaking it so as to get rid of the instabilities which are always there in the real world. Adaptive methods can actually reduce those costs and the need for ad hoc tweaking... but we still need safety methodologies to certify them and ways to use them more effectively. For example, we can learn how to use optimization methods to minimize the probability of a catastrophe, in situations where the probability cannot be realistically be brought down to zero no matter what we do.

Last year, a colleague at the Office of Naval Research said that the biggest need for new research, in his view, is the development of a new structure of V&V, different from the simple six sigma methods usually used for large-scale systems and different from the theorem-based approach used for simplified component subsystems. Perhaps this book will stimulate some of you to make contributions to that crucial emerging area.

August 2009

Paul Werbos
Washington, DC

The editors would like to thank Hamed Jafari for careful work on formatting and typesetting. The editorial work was in part funded by NASA Aeronautics, Separation Assurance Program.

Contents

Erratum

List of Contributors

Anuradha M. Annaswamy
Department of Mechanical Engineering
Massachusetts Institute of Technology
Cambridge, MA 02139.
aanna@mit.edu

Jim Barhorst
The Boeing Company.
jim.barhorst@boeing.com

Bojan Cukic
Lane Department of Computer Science
and Electrical Engineering
West Virginia University
Morgantown, WV 26506.
cukic@csee.wvu.edu

Dragan Djurdjanovic
Department of Mechanical Engineering
University of Texas at Austin
1 University Station,
C2200 Austin, TX 78712.
dragand@me.utexas.edu

Edgar Fuller
Department of Mathematics West Virginia University Morgantown, WV 26506.
ef@math.wvu.edu

Medorian D. Gheorghiu
North Carolina A&T State University
1601 East Market St.
Greensboro, NC, 27410.
medorian@yahoo.com

Pramod Gupta
University Affiliated Research Center
NASA Ames Research Center
Mail Stop 210-1
Moffett Field, CA 94035.
Pramod.Gupta-1@nasa.gov

Stephen A. Jacklin
NASA Ames Research Center
Mail Stop 269-2
Moffett Field, CA 94035.
Stephen.A.Jacklin@nasa.gov

Jinho Jang
Department of Mechanical Engineering
Massachusetts Institute of Technology
Cambridge, MA 02139.
jinho@mit.edu

Eugene Lavretsky
The Boeing Company
Huntington Beach, CA 92647.
eugene.lavretsky@boeing.com

Xiaoou Li
Departamento de Computacion
Mexico D.F., Mexico 07360.
lixo@cs.cinvestav.mx

Jianbo Liu
GM Research and
Development Center
30500 Mound Rd.
Warren, MI 48090.
jianbo.liu@gm.com

Yan Liu
Motorola Labs Motorola Inc.
Schaumburg, IL 60196.
yanliu@motorola.com

Kenneth A. Marko
ETAS Inc. 3021 Miller Rd.
Ann Arbor, MI 48103.
ken.marko@etas.com

Ali R. Mehrabian
Department of Electrical
and Computer Engineering
Concordia University
1455 de Maisonneuve Blvd. West
Montreal, Quebec,
Canada H3G1M8.
armehrabian@ieee.org

Mohammad B. Menhaj
Department of Electrical Engineering
Amirkabir University
of Technology
Hafez Ave.
Tehran, Iran 15914.
tmenhaj@ieee.org

Nhan T. Nguyen
NASA Ames Research Center
Mail Stop 269-2
Moffett Field, CA 94035.
Nhan.T.Nguyen@nasa.gov

Jun Ni
Department of Mechanical Engineering
University of Michigan
2350 Hayward St.
Ann Arbor, MI 48109.
junni@umich.edu

Alireza Rezazadeh
Department of Electrical Engineering
Shahid Beheshti University
Tehran, Iran 19839.
alireza.rezazade@gmail.com

Johann Schumann
RIACS/USRA,
NASA Ames Research Center
M/S 269-3
Moffett Field, CA 94035, USA
Johann.M.Schumann@nasa.gov

Mostafa Sedighizadeh
Department of Electrical Engineering
Shahid Beheshti University
Tehran, Iran 19839.
m_sedighi@sbu.ac.ir

Tim Smith
The Boeing Company.
timoth.a.smith4@boeing.com

David Song
Beijing Jiaotong University
College of Information Engineering
Beijing, China, 1000044.
songyd_ncat@yahoo.com

James M. Urnes
The Boeing Company.
james.m.urnes-sr@boeing.com

Liguo Weng
North Carolina
A&T State University
1601 East Market St.
Greensboro, NC, 27410.
Liguoweng@hotmail.com

Wen Yu
Departamento de Control Automatico
Mexico D.F., Mexico 07360.
yuw@ctrl.cinvestav.mx

Xiao-Hua Yu
California Polytechnic
State University
1 Grand Avenue
San Luis Obispo,
CA 93407, USA.
xhyu@calpoly.edu

Sampath Yerramalla
Controls and Engine Diagnostics
Pratt & Whitney, USA
400 Main St,
Mail Stop 163-14,
East Hartford, CT 06108, USA
syerramalla@gmail.com

Application of Neural Networks in High Assurance Systems: A Survey

Johann Schumann, Pramod Gupta, and Yan Liu

Abstract. Artificial Neural Networks (ANNs) are employed in many areas of industry such as pattern recognition, robotics, controls, medicine, and defence. Their learning and generalization capabilities make them highly desirable solutions for complex problems. However, they are commonly perceived as black boxes since their behavior is typically scattered around its elements with little meaning to an observer. The primary concern in safety critical systems development and assurance is the identification and management of hazards. The application of neural networks in systems where their failure can result in loss of life or property must be backed up with techniques to minimize these undesirable effects. Furthermore, to meet the requirements of many statutory bodies such as FAA, such a system must be certified. There is a growing concern in validation of such learning paradigms as continual changes induce uncertainty that limits the applicability of conventional validation techniques to assure a reliable system performance. In this paper, we survey the application of neural networks in high assurance systems that have emerged in various fields, which include flight control, chemical engineering, power plants, automotive control, medical systems, and other systems that require autonomy. More importantly, we provide an overview of assurance issues and challenges with the neural network model based control scheme. Methods and approaches that have been proposed to validate the performance of the neural networks are outlined and discussed after a comparative examination.

1 Introduction

Since the 1980s, artificial neural networks have evolved from biologically inspired layered networks connected by neurons into various categories of computational models with different algorithms and a large variety of architectural designs. They have been widely adopted in many applications as

J. Schumann, Y. Liu (Eds.): Appl. of Neural Networks in High Assur. Sys., SCI 268, pp. 1–19.
springerlink.com

function approximators and classifiers for over 30 years. Typical applications include function approximation, pattern recognition, image recognition, AI applications, and control. Over the past decades, applications of neural networks have emerged in many industrial fields, performing challenging tasks in medical experiments, flight control, automotive industry, robotics, power plants, etc. In many of these applications, neural networks have outperformed traditional computational approaches, with their compelling adaptive capabilities in learning and reacting to changing environments to accommodate novel situations, noise, and even failures. For systems that often operate in an evolving environment and thus requires a high level of self-adaptation, the employment of neural networks for online adaptation to accommodate system faults and recuperate against environmental changes has revolutionized the operation of real-time automation and control applications. However, the majority of neural networks are commonly treated as black boxes and their performance heavily relies on empirical validation. Since many of these industrial applications are in high assurance areas, the lack of in-depth information about the neural network learning coupled with its inherent nonlinearity has posed a great challenge for high assurance applications.

High assurance systems are those that require a warranted high level of robustness in system performance and a guaranteed set of critical properties including reliability, availability, safety, security, as well as other important qualitative attributes. Convincing evidence of adequately meeting the high assurance requirements must be provided when these systems are used in environments where failure can cause substantial financial loss or even loss of life. Hence, without a doubt, high assurance systems are of extreme importance in high profile missions, safety critical operations, and military applications. Examples include avionics, space exploration missions, power plant control, weapon control, life support in medical applications, etc.

Research communities, avionics in particular, have addressed the need for high assurance systems as well as the challenging issues they entail in terms of system verification, validation, and certification. The United States Federal Aviation Administration (FAA) has issued a series of certification processes for quality assurance of software used in aircraft (most notably the DO-178B standard (8)), which heavily rely on software verification and validation methods and theories. Verification and validation (V&V) is considered a crucial path towards certification in order to obtain proven reliability, availability, and safety for high assurance systems. For traditional software systems that are designed, developed, and deployed according to the common software life-cycle and well certified development models, empirical testing is considered an imperative and quite effective means to verify and validate a software. These tests take various forms such as coverage testing, requirements testing, regression testing, stress testing, etc. However, testing of safety-related software has been found to be a main driver for software costs (e.g., (10)).

The adoption of neural networks in an increasing number of high assurance systems has raised growing concern regarding their performance assurance.

Most traditional V&V methods are rendered inapplicable and/or ineffective for neural networks. This is simply because, unlike any traditional software, the structure of these novel computational models can evolve over time; and as a consequence, their performance, even validated, would remain subject to unforeseen changes, potentially causing instability and uncertainty in system operations. One of the most critical application areas for neural networks is control. The wide employment of neural networks in control started in 1990's with the advances of adaptive control theory (e.g., (3; 50)) and the rise of dynamic learning models. Many of these control applications are high assurance systems as they are designed/required to tolerate system failures and respond to these failures promptly in order to continue operation and preserve system stability and integrity. For example, as one of the most promising real-time automation and control applications, an adaptive flight control system must be designed to achieve adaptability through judicious online learning, aiding the adaptive controller to recover from operational damage (sensor/actuator failure, changed aircraft dynamics due to broken aileron or stabilator, etc.). Some of these conditions are severe enough to be considered "failure" mode conditions that significantly affect system performance. The National Aeronautics and Space Administration (NASA) conducted a series of experiments evaluating adaptive computational paradigms (neural networks, AI planners) for providing fault tolerance capabilities in flight control systems following sensor and/or actuator failures[1]. Experimental success suggests significant potential for further development and deployment of adaptive controllers (24; 5).

In this paper, we survey the application of neural networks in high assurance systems that have emerged in various fields, with a strong focus on control applications which include flight control, chemical engineering, power plants, automobile control, medical systems, and other systems that require autonomy. Section 2 provides a comprehensive list of neural network applications in these domains. Section 3 presents an overview of the assurance issues and challenges with the neural network model based control scheme. Methods and approaches that have been proposed to validate the performance of the neural networks are also outlined and discussed after a comparative examination. Section 4 concludes the paper with a summary and some observations of how future research could improve upon the existing body of applications and studies of neural networks in high assurance systems.

2 Application Domains

In the following, we will have a closer look at several important industrial areas, where neural networks have been applied. This selection is certainly

[1] For the NASA IFCS (Intelligent Flight Control System) project see `http://www.nasa.gov/centers/dryden/research/IFCS/index.html` and several chapters in this book.

not complete (for other areas, e.g., business see (29)) and it puts some focus on the application of neural networks for control. Other literature surveys include (28; 30), or (9).

2.1 Aircraft Control

One of the most prominent applications of neural networks is its use in damage-adaptive aircraft control. NASA's IFCS project (as discussed earlier) has been able to demonstrate that a manned aircraft can be successfully controlled using a neural network in the presence of (simulated) damage to the aircraft that change its aerodynamic behavior. Details will be presented in several chapters of this book. There are many different approaches to network-based adaptive control for aircraft, e.g., (39) for autonomous helicopters or (7). Different control architectures as well as different kinds of neural networks (e.g., Multilayer Perceptrons (MLP), Sigma Pi, Adaline networks, Radial Basis Functions (RBF), or Dynamic Cell Structures (DCS)) have been used for this task. In some approaches, the neural network will actually try to learn the dynamics model of the damaged aircraft; other approaches rely on control-augmentation, i.e., the neural network is trained to produce an additional control signal to counteract the effects of the damage. Very tight timing requirements set this application of neural networks aside from other applications: the neural network training algorithm must be capable to adjust within a few seconds. Otherwise, the damaged aircraft can get into an unrecoverable unstable state and will crash.

In this book, individual chapters will discuss important aspects of these kind of applications ranging from a more controls theoretic perspective to results and lessons learnt of the practical project with flight tests. Because of the high safety-criticality of aircraft control systems (in particular for manned aircraft), V&V is of particular importance and will be discussed in this book.

2.2 Automotive

With the advent of digital engine and drive-train control, neural networks have been used in this area, for example, to adapt toward different fuel quality or different driving styles. (20) describes an early drive-train control, which uses a neural network. More recently, advanced technology is being used to reduce fuel consumption, e.g., with a neural network based controller for the Toyota Prius (43), or to improve environmental impact by, e.g., network-based recognition of misfiring in diesel engines (12). Most applications in this area (for an overview see, e.g., (17)) are somewhat safety-critical and are supposed to work reliably in a wide range of situations. Factory recalls, due to software problems can be extremely costly, so a substantial effort is made toward verification and validation. In our book, we have two chapters devoted to neural networks in automotive applications. Other applications are concerned with the effective production of cars, e.g., by using neural networks to optimize the welding process (37).

2.3 Power Systems

The electric power industry is central for each country, as it has to reliably provide electric power to the customers, facing vastly increasing demands, aging infrastructure, and unforeseen natural events (e.g., lightning strikes, down power lines due to icing). Neural networks have been applied to various problems in design and development of power systems ever since the 1990's. Existing bibliographical studies of neural network applications to power systems outline five main application areas for neural networks in system operation and control: load forecasting, fault diagnosis, economic dispatch, security assessment, and transient stability. As a result of this literature review, a classification of publications on neural network applications to power systems between 1992 and 2004 is presented by Bansal (4). This article further discusses trends of adoption of this technology. Another classification of publications can be found in (18) for neural network applications to power systems (covering 2000 to 2005). Both analyses clearly show the growing interest as well as the success in applying neural networks to solve various problems in this domain. It is also noted that although the adoption has been popular, the variety of used neural network architecture remains limited: multi-layer perceptrons, Hopfield networks, and Kohonen neural networks are the three major models that have been employed in power systems.

Neural network applications in power systems have been well studied with a strong focus to improve its prediction accuracy. The economic value of employing neural networks for prediction, load forecasting, and economic dispatching in particular, is reinforced by their successful applications in areas like short-term electric demand forecasting or combustion optimization with reduced NOx emissions. Two major conferences were held in late 90's, the 1998 American Power Conference (1) and the 1999 International Business Forecasting Conference, featuring the findings and practices of neural network applications in these areas. An exemplary application of neural networks to nuclear power systems with an assessment of economic benefits is given by Lisboa (31). Neural networks are applied together with an expert system in a staged approach to retrieve useful information from a gigantic amount of data generated from inspections of the reactor core control assemblies (RCCAs). Approximately 800MBytes of data is produced per core inspection. By extracting out the 5% of data that contains the meaningful information using a neural network based approach, the inspection time is greatly reduced and as a result, the Duke Power Company could save a substantial cost estimated at $28,000 per inspection and projected to save "$361k in the next 5 years"[2].

[2] `http://www.nuc.berkeley.edu/thyd/ne161/rtse/dukerods.html`

In addition to forecasting, system control and operations also benefit from neural network applications. In power generation, neural networks can be used to estimate certain process variables to model a dynamic and often nonlinear process (e.g., pulverized fuel flow to the boiler) that otherwise cannot be measured or computed directly. Unlike linear modeling techniques (e.g., Kalman Filters), neural networks can approximate linear functions as well as any nonlinear functions and provide an accuracy at any required level with high cost efficiency. For this reason, more and more neural networks are adopted by power stations in dynamic environments for detecting and diagnosing faults and gaining transient stability. Neural networks are also used in the control and monitoring of steam turbines, as their proper operation strongly influences the overall power plant efficiency. Siemens, for example, is using neural networks for the estimation of the blade temperature (49; 36).

2.4 Medical Systems

It is obvious that many medical software applications are highly safety-critical; failures in the software can cost human lives. A prominent example is the Therac-25, a radiation therapy device, where faulty software caused several (fatal) accidents (38). The noisy and statistical nature of most medical data and measurements seem to be ideally suited for the application of neural networks. The earliest and most widely used neural network based system in health care is Papnet, which has been developed by Neuromedical Systems, Inc. in the 90's. There is a number of studies that this software improves detection rates for cervical cancer from Papnicolau stained smear slides. However, the cost-effectiveness of this application was never satisfactory (31).

An example of a hybrid decision support system in health care is GLADYS (GLAsgow system for the diagnosis of DYSpepsia)[3], developed by the Glasgow Southern General Hospital with support from the University of Glasgow's Department of Public Health. GLADYS uses a Bayesian model for the diagnosis of several conditions related to dyspepsia. It uses statistical representations to encode knowledge of clinical staff in a structural form that can be updated numerically and is used to process uncertain knowledge in a consistent manner.

Questar[4] (54) is a sleep analysis package, developed initially by the Engineering Department at Oxford University and marketed by Oxford Instruments. It was awarded a British Computer Society medal in 1996 and gained FDA approval in 1997. The purpose of this software is to automate sleep staging into awake, rapid eye movement (REM) or light sleep, and deep sleep as accurately as an expert user, but on a continuous scale and with a much faster sampling rate of 1Hz. It does this by combining three electrical

[3] `http://students/dcs.gla.ac.uk/students/lamkc/CPI.html`

[4] `http://www.eng.ox.ac.uk/World/Research/Summary/B-Neural.html`

measurements, electro-encephalogram (EEG), electro-oculogram (EOG) and electro-myogram (EMG), which measure mental activity, eye movement, and muscular activity, respectively.

Another health care project developed at Oxford University is a software monitor for intensive care patients. are very high. As a result, a The demand for intensive care beds is very variable and costs for intensive care are extremely high. Thus, the decision to whether or not to admit a critical patient into intensive care can have substantial impact. Such a decision can be supported with the aid of a statistical advisory system, e.g., the commonly used Apache II (Acute Physiology and Chronic Health Evaluation II, Glasgow University) (25),

In the same domain of management and intensive care for critically ill patients, several software packages using rigorous statistical methods and neural networks for knowledge discovery are used in European hospitals. A Bayesian model of clinical data has been used to test the hypothesis that Cerebral Partial Pressure does indicate the presence of sub-clinical damage by trending during the first 24 hours following admission (31). This indicates that careful monitoring of this highly invasive measurement can improve the management of patients, who "talk and die" (35).

The original Apache II monitor processes five standard physiological measurements (EEG, systolic blood pressure and oxygen saturation, breathing rate, and temperature) and produces alarms based on novelty of the data. Thus, it is an example of a data based system where the available signals define a nominal state, which is not of interest. The challenge is to accurately and robustly determine deviations in this multivariate data stream. The problem becomes harder due to the low density in the input space and the necessity to accommodate different states of the patient during recovery and robustness against artifacts (e.g., sensor displacement).

A web-based advisory system using neural networks has been developed for the automated interpretation of myocardial perfusion images[5]. Another system, also developed at Lund University, is used for acute myocardial infarction (AMI) detection. It was tested on a data base of approximately 1,000 electro cardiograms (ECGs) from patients with AMI and approximately 10,000 control ECGs. In this application, the neural network system was found to be more sensitive and has a higher discrimination accuracy than benchmark ECG software, or expert cardiologists.

2.5 Other Applications

The Sharp LogiCook (54) was the first microwave oven that used neural network technology. It was originally developed at Oxford University[6]. Based

[5] `http://www.weaidu.com/software/index.html`

[6] `http://www.eng.ox.ac.uk/World/Research/Summary/B-Neural.html` or `http://www.scit.wlv.ac.uk/~cm1822/acnl7.htm`

upon user input (food or liquids), the optimum cooking time was obtained from an analysis of the proportional, integral and derivative humidity profiles using a neural network. The software is also capable of dealing with different-sized portions and can detect dangerous conditions.

An industrial area where neural network control has been successfully applied for a long time is a steel rolling mill. Here, accurate control of temperature of the strip and the rolling force are critical for the quality of the product. Based upon a prototype, developed for Hoesch (Dortmund, Germany), Siemens has deployed this technology world wide since then. (45) claims efficiency gains of 30% due to better accuracy in rolling force modeling with prediction improvements leading to savings of $200K in material costs annually. In this application, the neural network's capability to handle non-linear data has been beneficial.

In the Airline business operation area, BehavHeuristics, Inc. (started in 1986 and later part of Airline Automation Inc.) uses reinforcement learning to predict no-shows in air-flights, thus maximizing the passenger load through controlled overbooking. Their Airline Marketing Tactician (AMT) (22) was an early success for neural networks.

The Boeing Company's NIRS (Neural Information Retrieval System) (48; 23), is probably still the largest scale manufacturing application of neural networks. It uses a binary Adaptive Resonance Theory network (ART1), to cluster binary templates of aircraft parts. The systems arranges them in a complex hierarchical network covering over 100,000 items. These are then grouped into thousands of self-organized clusters. Substantial savings in manufacturing costs (several $M per year) have been reported.

3 Toward V&V of NNs in High Assurance Systems

3.1 V&V of Software Systems

Any subsystem and component of a high assurance application must undergo a rigorous process in order to make sure that all requirements regarding safety, performance, and reliability are met. This refers to any hardware component as well as to software. Since most neural network based applications are ultimately implemented as a software program, we will focus on software components only.

In any software development lifecycle, there are, in addition to activities for the design and implementation of the software, activities to ensure that the final software is working as expected. Traditionally, we distinguish between verification and validation: Verification is often informally described as "Are you building the right thing?", whereas validation can be paraphrased as "Are you building the thing right?". It is also obvious that different V&V activities are performed at different stages of the software lifecycle. Figure 1 shows a simplified version of the software development stages and related

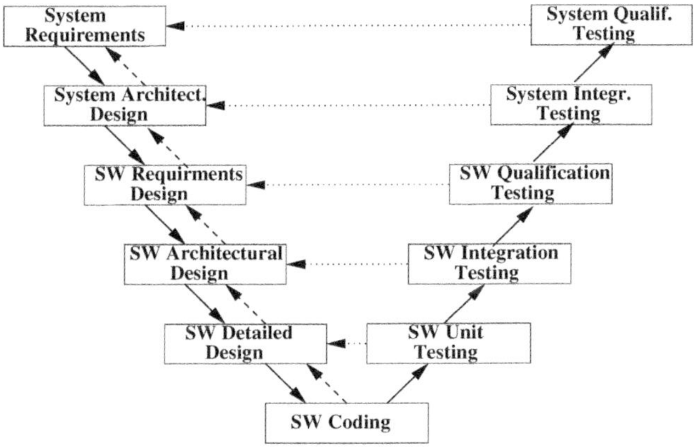

Fig. 1 Verification and validation activities during software development ("V-shape"). Verification activities are marked by dashed lines, validation by dotted arrows

V&V activities (see e.g., (42)). Several observations can be made, which will be helpful to tackle V&V issues for network based systems:

- Verification tasks are performed on the left side of the "V" and thus are mainly performed during earlier stages of the development process, whereas validation tasks (mainly testing) relate the finished products (on the right side of the "V") with the corresponding artifact on the left side. Ultimately, in the system qualification (or acceptance) testing, the entire system is to be tested against the requirement specifications.
- It is well known that the removal of faults can be orders of magnitude more expensive in later stages than in an early development phase (42). In particular for safety-critical applications, costs for V&V are the main cost drivers for software (e.g., (10)).
- Verification activities can be loosely grouped into design-time V&V and code V&V. In particular, when complex algorithms like neural network learning algorithms or multivariate optimization algorithms are used, this distinction is important and we will discuss it in detail below.
- Traditional V&V ends when the software is deployed. However, if the software is to work in unknown or changing environments, or has to react toward unforeseen events, additional activities are necessary in order to ensure that the software is working correctly. Such techniques range from simple exception handling and dynamic performance monitoring to run-time verification and certification (e.g., (46). Techniques for recovery from failures include reconfiguration or code repair (e.g., via self healing code (13)).

Virtually all software for high assurance applications is developed according to a specific software process. Usually such processes are highly standardized (e.g., according to IEEE or ANSI). (2) gives an overview of several traditional ANSI standards. Usually, a Software V&V process is an integral component of a software development process. A V&V process describes which tests are to be carried out, which activities for verification are to be performed, and how the tests and their results are documented.

In many safety-critical application areas, all systems (and also the software) have to go through a certification process. This often highly standardized process has the goal to demonstrate to a certification board that all required steps have been carried out and that due diligence has been applied to make sure that the system under consideration adheres to all safety and performance requirements.

Probably the best-known certification standard is DO-178B (8), which is the standard prescribed by the FAA for all safety-critical software to be used in civil transport aircraft in the US. It is a very detail-oriented and resource-consuming process, so certification is a major cost driver for safety-critical software.

As discussed earlier, existing standards, however, cannot be used as is for the V&V of neural networks. In the following, we will discuss a number of V&V issues, which prevent the use of current certification standards for neural network applications.

3.2 V&V Issues and Gaps for NN-Based Applications

Different scientific and engineering communities use different notations and nomenclature. This can lead to substantial misunderstandings like the following examples:

"non-deterministic". In computer science (CS), the notion of non-deterministic piece of code is always attached to a program "execution with one or more choice points where multiple different continuations are possible without any specification of which one will be taken"[7] . Practical implementations of non-determinism thus usually use random number generators. In general, a specific state (or computation sequence) cannot be reproduced, making testing of such software extremely difficult.

In engineering disciplines, a system is usually coined non-deterministic if it is non-Markovian, i.e., that the system state $\mathbf{x}_t$ cannot be totally described by $\mathbf{x}_{t-1}$. Rather, the entire history (e.g., the entire flight since take-off) needs to be taken into account, i.e., $\mathbf{x}_t$ can only be calculated given $\mathbf{x}_0, \mathbf{x}_1, \ldots, \mathbf{x}_{t-1}$. With all the history present, the state $\mathbf{x}_t$ can be exactly reproduced. Thus, this notion is not based upon any random number generators.

Virtually all adaptive control systems are non-Markovian, but deterministic (in the CS sense). Since some forms of neural network algorithms start with randomly initialized weights (e.g., standard multi-layer

[7] `http://en.wikipedia.org/wiki/Non-deterministic_algorithm`

perceptrons), the opinion that neural networks (and thus neural-networks based adaptive controllers) are nondeterministic (in the CS sense) persisted.

"Neural Network". Many architectures for adaptive control systems have been developed using neural networks (e.g., (40; 47; 7)). The research area of neural networks, in general, has traditionally been put into the vicinity of artificial intelligence. Hence, the notion of a neural network is often attached to terms like "AI", "brain-like", "bio-inspired" possibly leading to confusion and low confidence, when considered within a safety-critical environment.

Technically speaking, the neural networks in the adaptive controllers have been purely used as multivariate non-linear function approximators; the "learning" is (in most cases) a recursive least-squares optimization algorithm. Described in these terms, a lot of "hype" about potential and "dangers" of neural networks can be avoided.

An adaptive control system or other NN based software in a high assurance application is handled like any other highly safety-critical piece of software: it must undergo rigorous V&V and the software must be certified. However, most traditional techniques for V&V as prescribed in these standards cannot be used on an online neuro-adaptive system because this system

- has to deal with a dynamically changing, unknown, non-linear plant model. Typically, damages to an aircraft (e.g., a stuck rudder) introduces biases, non-linearities and unknown interactions (e.g., correlations between the different aircraft axes). Moreover, most aerospace analysis techniques are restricted to the linear case.
- is a system, which contains non-linear functions and approximators. Except for the most primitive kinds of neural networks, neural networks use nonlinear activations functions and can, in principle, approximate any smooth function.
- is adapted using a complicated algorithm. In most cases, the neural network is being trained using some kind of machine learning algorithm. Such algorithms usually are variants of a recursive multivariate (quadratic) numerical optimization routine.

3.3 V&V Approaches for Neural Networks

In the following, we will discuss V&V approaches for neural networks and systems, containing neural networks, in particular neuro-adaptive controllers. This area can be roughly subdivided into the following categories, concerning techniques that

- specifically subdivide V&V activities into algorithm V&V and code V&V. In particular, theoretical results, obtained during algorithm V&V (e.g., Lyapunov stability proofs) must be used to guide and augment code V&V.

- focus on the analysis of the neural network architectural design (e.g., number of hidden layers, number of hidden nodes),
- consider neural networks as function approximators or data classification tools,
- help the human reader to understand the inner workings of the neural network (e.g., by rule-extraction or representation as a Fuzzy System),
- focus on the specifics and characteristics of the learning (training) algorithm,
- analyze the selection and quality of the data used for adaptation, and
- provide means for the dynamic (i.e., during operation) monitoring of the performance of the adaptive component.

Obviously, the techniques and approaches in each of these categories heavily overlap and have synergistic effects. Moreover, the various techniques range from mathematical theorem and proof (e.g., universal function approximation of a MLP), statistical methods, methods from design of experiments, testing, simulation, and dynamic analysis and monitoring of the behavior and performance of the neural network. The term "dynamic" here indicates that the monitoring occurs during the actual operation of the neural network based system after deployment.

In many cases, certain performance and safety aspects of the neural network are necessary in order to analyze the larger system. A typical example is a neural-networks based adaptive controller. In order to show (eventual) stability of the controller using Lyapunov stability theory, assumptions about the neural network (e.g., on bounds of the error) are required. Such proofs can be pretty involved. In several chapters of this book such stability proofs are discussed. Other examples can be found, for example, in (47).

3.3.1 NN as Function Approximator

Traditional literature describes adaptive computational paradigms, neural networks in particular, with respect to their use, as function approximators or data classification tools. Validation on these systems is usually based on a train-test-re-train empirical procedure. Some bibliographic references also propose methods as part of the training algorithm of neural networks for validation (55; 6). The ability of interpolating and/or extrapolating between known function values is measured by certain parameters through testing. This evaluation paradigm can be reasonably effective only for pre-trained adaptive systems, which does not require online learning and adaptation and remain unchanged in use.

3.3.2 V&V for NN Design

In (11), Fu interprets the verification of a neural network to refer to its correctness and interprets the validation to refer to its accuracy and efficiency. He establishes correctness by analyzing the process of designing the neural network, rather than the functional properties of the final product. Peterson presents another similar approach in (41) by discussing the software development process of a neural network. He describes the opportunities for verification and validation of neural networks in terms of the activities in their development life cycle, as shown in Figure 2.

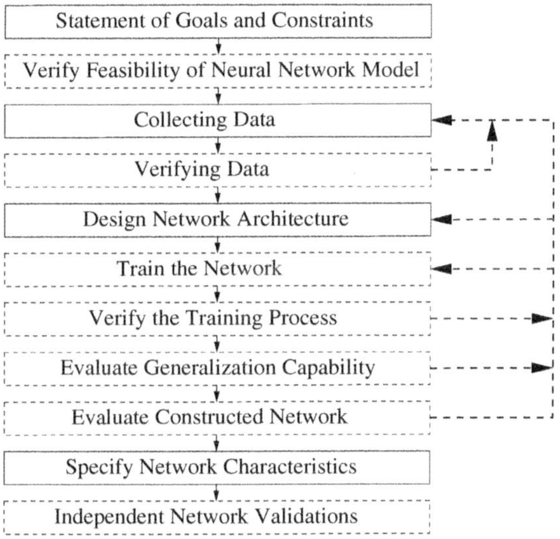

Fig. 2 The development cycle of a neural network

As we can see from Figure 2, there is a focus on V&V of adaptive systems based on the training data. Verification of the training data includes the analysis of appropriateness and comprehensiveness. However, in online learning mode, this technique may not be appropriate due to its real-time training aspects. Data are collected in such a way that the training is completed under intensive computational requirements. An applicable approach for verifying the data is novelty detection.

3.3.3 V&V for NN Training

Verification of the training process typically examines the convergence properties of the learning algorithm, which is usually pre-defined by some criteria of error measure. In (21), K.J. Hunt et. al. investigate all different methods for error estimation techniques and make detailed comparison among them. Nonetheless, effective evaluation methods of interpolation and extrapolation capabilities of the network and domain specific verification activities are still based on empirical testing (26). Literature addressing the problem analytically is very scarce. In the field of function approximation theory, MLP networks have been proven to be universal approximators for being able to achieve any given accuracy provided a sufficient number of hidden neurons (19). The mathematical analysis and proof can be seen as another effort for validating the learning process as it can provide theoretical proof for the capabilities of function approximation. The weakness of such analytical proof is that it remains impractical for online adaptive learning systems as the system function evolves.

Most recently proposed techniques on V&V of neural networks are based on empirical evaluation through simulation and/or experimental testing using statistical methods such as K-fold cross-validation, bootstrapping, repeated random sampling, etc. There also exists some approaches to V&V of dynamic neural networks by modifying the training algorithms. In an attempt to solve the dilemma of plasticity and stability for neural networks, S. Grossberg (14; 15) derives a new paradigm, referred to as the Adaptive Resonance Theory (ART-1/2/3). Within such a network, there are two components charging seen and unseen data respectively. The Validity Index network presented by Leonard et. al. in (27) is an example of modification to the network training algorithm for V&V of the neural networks. When tested, the validity index in a Radial Basis Function neural network provides a confidence interval associated with each network prediction for a given input.

3.3.4 Dynamic Monitoring

For online neural networks that are adopted in adaptive control applications, static V&V methods for neural network design and training fall short to warrant online performance assurance due to the dynamic nature of the network. Because not all conceivable situations can be validated upfront, it is almost impossible to guarantee the assurance of reliable performance and safety. In order to validate the online adaptation performance, dynamic monitoring tools can be used that work during the actual execution of the software. Such tools that can dynamically monitor the quality of the neural network and its internal parameters have been proposed, mainly focusing on the learning performance and prediction performance. A few major approaches are listed below.

- Online Learning Performance Analysis.
  Lyapunov stability theory based monitors are proposed for the Dynamic Cell Structure (DCS) Networks in Yerramalla et. al. (56; 57) and Chapter 6 of this book. The proposed online monitoring system is composed of several dynamic stability monitors. Each monitor is essentially a Lyapunov-like function that is designed to analyze and capture unstable behavior from a particular aspect of online learning. These monitors provide an observation of how well a set of associated neural centers of the online neural network are being overlaid over corresponding relative elements of the presented training data set.

  Schumann and Liu (52; 53) propose another estimate for online learning performance by calculating the *parameter sensitivity* in the context of flight controllers. In an online adaptive system, the internal control parameters are changing while the system is in operation. The sensitivity of a parameter with respect to changes is computed as the probability $p(\mathbf{o}|\mathcal{P}, \mathbf{x})$ for network output $\mathbf{o}$, given parameters $\mathcal{P}$, and inputs $\mathbf{x}$. Assuming a Gaussian probability distribution, the parameter sensitivity can be obtained as the variance $\sigma^2_{\mathcal{P}}$. Such measures can provide useful information to improve neural network design and learning paradigms.

- Network Prediction Confidence Estimation.
The Validity Index tool proposed by Liu et. al. (32; 34) calculates reliability measures for the output of a DCS network, which has been used for the Gen-I IFCS controller. Following the definition of Validity Index (VI) in RBF networks by Leonard et. al.(27), the validity index in DCS networks is defined as an estimated confidence measure of a DCS output, given the current input. The VI can be used to measure the accuracy of the DCS network fitting and thus provides information for future validation activities. By examining the statistical properties of the best matching neuron and its neighbors that are activated during learning and prediction, the validity index tool takes into account the topology-based learning structures and produces a quality metric for the output. Details can be found in (33).
Schumann et. al. (51) developed the Confidence Tool using a Bayesian statistical approach to estimate the quality of learning and the accuracy of estimation. Considering all inputs and outputs of the neural network as statistical variables with a given probability density function (e.g., Gaussians), the algorithm determines the variance σ^2 as a quality metric on the output. This tool has been developed for the IFCS adaptive flight controller and has been test-flown on a manned F-15 NASA aircraft.

4 Conclusions

This chapter serves as our attempt to provide an overview of applications of neural networks, where failure is not an option. Such high-assurance applications can be found in many domains, most prominently in aerospace, automotive industry, medical applications, and power industry. Here, the use of neural networks provides substantial benefits with respect to performance, accuracy, and/or handling of unforeseen situations. However, the algorithms implementing neural networks can be very complex, in particular the training algorithms, which adjust the neural network's parameters based upon given data. These training algorithms, most often variants of non-linear multivariate quadratic optimization algorithms, are at the core of many neural network based applications. Because these applications, which have been discussed above, are safety-critical i.e., failures or erroneous behavior can ultimately claim human lives, the software that implements the algorithmic learning of neural networks has to undergo rigorous verification and validation (V&V) before deployment. In this chapter we discussed that traditional V&V for safety-critical code is not sufficient for neural-network based applications, in particular for those applications where the neural network is trained in an online fashion during operation.

The presented overview also serves the purpose of structuring the field, as well as illustrating the widespread application potential of neural networks in safety-critical applications and the issues in terms of their V&V that had to be addressed by the research community.

The remaining chapters of this book are ordered in a similar way. Chapters 2 and 3 discuss theoretical and design-time analysis on semi-global boundedness and margins of adaptive control and assessment of network complexity.

The next several chapters are devoted to applications of neural networks for adaptive aircraft control. One chapter focuses on the development of a damage-adaptive flight controller all the way through manned flight tests. The other one presents approaches for stability and convergence of adaptive flight control, a central challenge for V&V of such controllers. Some neural network architectures change their architecture and size, while they are being trained. Using such networks in an adaptive controller poses substantial challenges for V&V. This chapter deals with dynamic allocation in such neural network architectures.

The next chapter is centered around the automobile: it describes the use of an immune-system approach to help with the localization of faults in automotive engines. Moving yet to another element, the subsequent chapter discusses the design of a neuro-adaptive controller for a submarine.

In most mechanical systems, friction between moving components is a major issue. Our chapter describes how neural networks can be used to provide accurate friction control. Due to the transition between slipping and sticking this problem is highly nonlinear. The final two chapters discuss how neural networks can improve the efficiency of processes (blending of crude oil) and fuel cells.

We hope that the wide range of applications and methods described in the book illustrate the potential of neural networks in safety-critical and high-assurance applications and help the reader to be more aware of issues and approaches and to drive the advances of V&V of such systems to ultimately make them safe and reliable.

References

1. American Power Conference. Proceesings of the American Power Conference 1998, vol. 60, 2 vol. set (1998)
2. American National Standards. Software Engineering Standards. IEEE, Wiley and Sons (1984)
3. Astrom, K.J., Wittenmark, B.: Adaptive Control. Addison-Wesley, Reading (1994)

4. Bansal, R.C.: Overview and literature survey of artificial neural networks applications to power systems (1992–2004). Journal - Institution of Engineers India. Electrical Engineering Division (86), 282–296 (2006)
5. The Boeing Company. Intelligent Flight Control: Advanced Concept Program. Technical Report (1999)
6. Boyd, M.A., Schumann, J., Brat, G., Giannakopoulou, D., Cukic, B., Mili, A.: Validation and verification process guide for software and neural nets. Technical report, NASA Ames Research Center (2001)
7. Calise, A., Rysdyk, R.: Nonlinear adaptive flight control using neural networks. IEEE Control Systems Magazine 21(6), 14–26 (1998)
8. DO-178B: Software Considerations in Airborne Systems and Equipment Certification (1992), `http://www.rtca.org`
9. DTI. DTI final report: Evaluation of parallel processing and neural computing application programmes. Assessment Paper 38 URN 99/922 (1999)
10. Ebert, C., Parro, C.H., Suttels, R., Kolarczyk, H.: Improving Validation Activities in a Global Software Development. In: Proc. ICSE 2001, p. 545. IEEE, Los Alamitos (2001)
11. Fu, L.: Neural Networks in Computer Intelligence. McGraw Hill, New York (1994)
12. Getman, A., Zhou, H., Jammalamadaka, S.: An automated network for detecting diesel engine misfire. In: Proc. IJCNN 2007, pp. 3017–3021. IEEE, Los Alamitos (2007)
13. Ghosh, D., Sharman, R., Rao, R.H., Upadhyaya, S.: Self-healing systems — survey and synthesis. Decision Support Systems 42(4), 2164–2185 (2007)
14. Grossberg, S.: Adaptive pattern classification and universal recoding: I. Parallel development and coding of neural feature detectors. Biological Cybernetics 23, 121–134 (1976); Reprinted in Anderson and Rosenfeld (1988)
15. Grossberg, S.: Competitive learning: From interactive activation to adaptive resonance. Cognitive Science 11(1), 23–63 (1987)
16. Gupta, P., Guenther, K., Hodgkinson, J., Jacklin, S., Richard, M., Schumann, J., Soares, F.: Performance Monitoring and Assessment of Neuro-Adaptive Controllers for Aerospace Applications Using a Bayesian Approach. In: Guidance, Navigation and Control (GNC) Conference, AIAA 2005-6451 (2005)
17. Gusikhin, O., Rychtyckyj, N., Filev, D.: Intelligent systems in the automotive industry: Applications and trends. Knowl. Inf. Syst. 12(2), 147–168 (2007)
18. Tarafdar Haque, M., Kashtiban, A.M.: Application of neural networks in power systems: A Review. Transaction on Engineering, Computing and Technology (6), 53–56 (2005)
19. Hornik, K.M., Stinchcombe, M., White, H.: Multilayer Feedforward Networks are Universal Approximators. Neural Networks 2, 359–366 (1989)
20. Hrycej, T.: Neural-network-based car drive train control. In: Proc. 42nd IEEE Vehicular Technology Conference, vol. 2, pp. 1042–1045 (1992)
21. Hunt, K.J., Sbabaro, D., Zbikowski, R., Gawthrop, P.J.: Neural Networks for Control Systems—A Survey. Automatica 28(6), 1707–1712 (1996)
22. Hutchison, W.R., Stephens, K.R.: The airline marketing tactician (AMT): A commercial application of adaptive networking. In: Proc. First IEEE International Conference on Neural Networks, vol. 2, pp. 753–756. IEEE, Los Alamitos (1987)

23. Irwin, D.: The industrial Electronics Handbook. Technology & Engineering (1997)
24. Jorgensen, C.: Feedback linearized aircraft control using dynamic cell structures. In: World Automation Congress (ISSCI), Alaska, pp. 050.1-050.6 (1991)
25. Knaus, W.A., Draper, E.A., Wagner, D.P., Zimmerman, J.E.: Apache II: a Severity of Disease Classification System. Critical Care Medicine 13, 813–829 (1985)
26. Lawrence, S., Tsoi, A.C., Back, A.D.: Function approximation with neural networks and local methods: Bias, variance and smoothness. In: Australian Conference on Neural Networks, pp. 16–21 (1996)
27. Leonard, J.A., Kramer, M.A., Ungar, L.H.: Using radial basis functions to approximate a function and its error bounds. IEEE Transactions on Neural Networks 3(4), 624–627 (1992)
28. Lisboa, P. (ed.): Current Applications of Neural Networks. Chapman and Hall, Boca Raton (1992)
29. Lisboa, P., Vellido, A., Edisbury, B. (eds.): Neural Network Applications in Business. World Scientific, Singapore (2000)
30. Lisboa, P., Ifeachfor, E., Szczepaniak, P. (eds.): Artifical Neural Networks in Biomedicine. Springer, Heidelberg (2000)
31. Lisboa, P.: Industrial use of safety-related artificial neural networks. Contract Research Report 327/2001, Liverpool John Moores University, Liverpool (2001)
32. Liu, Y., Yerramalla, S., Fuller, E., Cukic, B., Gururajan, S.: Adaptive Control Software: Can we guarantee safety? In: Proc. of the 28th International Computer Software and Applications Conference; Workshop on Software Cybernetics (2004)
33. Liu, Y., Cukic, B., Jiang, M., Xu, Z.: Predicting with Confidence—An Improved Dynamic Cell Structure. In: Advances in Neural Computation, vol. 1, pp. 750–759. Springer, Heidelberg (2005)
34. Liu, Y.: Validating A Neural Network-based Online Adaptive System. PhD thesis, West Virginia University, Morgantown (2005)
35. Marshall, L.F., Tool, B.M., Bowers, S.A.: The National Traumatic Coma Data Bank Part 2: Patients who talk and deteriorate: Implications for treatment. Journal of Neurosurgery 59(2) (1983)
36. Mathur, A., et al.: Turbine Back Pressure Identification and Optimization with Learning Neural Networks. Advances in Instrumentation and Control (45): 1, 229–236 (1990)
37. Ramírez Mendoza, R.A., Morales-Menéndez, R., Cantú-Ortiz, F.J.: Neural nets modelling for automotive welding process. In: Hamza, M.H. (ed.) Modelling and Simulation. Acta Press (2005)
38. Neumann, P.G.: Computer Related Risks. ACM Press, New York (1995)
39. Ng, A., Jin Kim, H.: Stable adaptive control with online learning. In: Proceedings of NIPS, vol. 17 (2005)
40. Norgaard, M., Ravn, O., Poulsen, N.K., Hansen, L.K.: Neural Networks for Modeling and Control of Dynamic Systems. Springer, Heidelberg (2002)
41. Peterson, G.E.: A foundation for neural network verification and validation. In: SPIE Science of Artificial Neural Networks II, vol. 1966, pp. 196–207 (1993)
42. Pressman, R.: Software Engineering: A Practitioner's Approach. McGraw-Hill, New York (1999)

43. Prokhorov, D.: Toyota Prius HEV Neurocontrol and Diagnostics. Neural Networks 21(2-3), 458–465 (2008)
44. Pullum, L., Taylor, B., Darrah, M.: Guidance for the Verification and Validation of Neural Networks. Emerging Technologies. Wiley-IEEE Computer Society Press (2007)
45. Ruha, P., Peura, P., Oyj, R., Jansen, M.: Using Neural Networks in Steel Strip Rolling. In: Steels and Materials for Power Plants, pp. 51–56. Wiley, Chichester (2006)
46. Rushby, J.: Runtime Certification. In: Leucker, M. (ed.) RV 2008. LNCS, vol. 5289, pp. 21–35. Springer, Heidelberg (2008)
47. Rysdyk, R., Calise, A.: Fault tolerant flight control via adaptive neural network augmentation. AIAA American Institute of Aeronautics and Astronautics, AIAA-98-4483, pp. 1722–1728 (1998)
48. Smith, S.D.G., Escobedo, R., Anderson, M., Caudell, T.P.: A deployed engineering design retrieval system using neural networks. IEEE Trans. on Neural Networks 8(4), 847–851 (1997)
49. Santoso, N., Petsche, T.: A Method for Blade Temperatrue Estimation in a Steam Turbine. European Patent, No EP0944866 (2001)
50. Sastry, A., Bodson, M.: Adaptive Control: Stability, Convergence and Robustness. Prentice Hall, Englewood Cliffs (1994), `http://www.ece.utah.edu/~bodson/acscr`
51. Schumann, J., Gupta, P.: Monitoring the performance of a neuro-adaptive controller. In: Proc. of the 24th Intl. Workshop on Bayesian Inference and Maximum Entropy Methods in Science and Engineering (2004)
52. Schumann, J., Liu, Y.: Performance Estimation of a Neural Network-based Controller. In: Wang, J., Yi, Z., Żurada, J.M., Lu, B.-L., Yin, H. (eds.) ISNN 2006. LNCS, vol. 3972, pp. 981–990. Springer, Heidelberg (2006)
53. Schumann, J., Liu, Y.: Tools and Methods for the Verification and Validation of Adaptive Aircraft Control Systems. In: IEEE Aerospace Conference. IEEE Press, Los Alamitos (2007)
54. Tarassenko, L.: LogiCook and QUESTAR: two case studies in successful technology transfer. IEE Digest 1997(14) (1997)
55. Tibshirani, R.: Bias, Variance and Prediction Error for Classification Rule. Technical Report, Statistics Department, University of Toronto (1996)
56. Yerramalla, S., Cukic, B., Fuller, E.: Lyapunov Stability Analysis of Quantization Error for DCS Neural Networks. In: Proc. Intl. Joint Conference on Neural Networks, IJCNN 2003 (2003)
57. Yerramalla, S., Fuller, E., Cukic, B.: Lyapunov Analysis of neural network stability in an adaptive flight control system. In: Huang, S.-T., Herman, T. (eds.) SSS 2003. LNCS, vol. 2704, pp. 77–91. Springer, Heidelberg (2003)

Robust Adaptive Control Revisited: Semi-global Boundedness and Margins

Anuradha M. Annaswamy, Jinho Jang, and Eugene Lavretsky

Abstract. While stability of adaptive control systems and their robustness to disturbances, unmodeled dynamics, and time-delays have been extensively analyzed in the literature, what is missing is the development of quantifiable margins that explicitly define the extent of these perturbations that the adaptive controller can withstand. In this paper, a first attempt is made to quantify robustness of adaptive controllers to time-delays, unmodeled dynamics, and state-dependent nonlinearities. Formal stability margin analysis is conducted and the theoretical predictions are validated using high performance aircraft flight simulation environment.

1 Introduction

Over the past three decades, a large body of literature has been developed for the control of complex systems in the presence of uncertainties. Adaptive systems have been extensively studied over past years and their stability properties have been derived in the presence of bounded unknown disturbances and unmodeled dynamics (1; 3; 5; 6; 7). However, it should be noted that theoretically verifiable Verification and Validation (V&V) techniques for adaptive systems are conspicuously absent. Current V&V techniques rely on the fact that the underlying control system is linear, which makes them inadequate for adaptive control systems which are intentionally nonlinear. Currently, the chief practical obstacle to transitioning adaptive controllers into aerospace applications is an inability to analytically assert that the closed-loop system will have adequate stability/robustness margins with respect to perturbations such as time-delays, unmodeled dynamics, and disturbances. In this paper, we derive such margins.

The starting point for the derivation of these margins is the well known stability properties of adaptive systems. In the absence of perturbations due to unmodeled dynamics, delays, nonlinearities, or disturbances, the underlying closed-loop system can be shown to remain stable under certain conditions

J. Schumann, Y. Liu (Eds.): Appl. of Neural Networks in High Assur. Sys., SCI 268, pp. 21–39.
springerlink.com

(6). In such cases, the closed-loop system transitions to a linear time-invariant system (in the case of constant command signals) asymptotically. Therefore, one can argue that stability margins of the adaptive system tend toward this "asymptotic" linear system. However, the response of the adaptive system, understandably, is more vulnerable to perturbations during the initial transient stage during which time the overall system is markedly nonlinear. As such, the degrees of robustness that the adaptive system has with respect to the perturbations during the transient stage are the margins that become relevant. Therefore, in this paper, we focus on the margins of the adaptive flight control system (AFCS) during the adaptive phase. Standard stability and robustness tools used for adaptive systems in (3; 6) are utilized in order to derive these margins. Problem statement, adaptive controller, stability margins, and validation in the context of a high performance aircraft are included in the subsequent sections. In order to elucidate the approach used, we focus our discussions on a single-input system. The results can be extended in a straightforward manner to multi-input systems as well.

2 Problem Statement

The motivation for adaptive control stems from several causes including aerodynamic uncertainties, modeling inaccuracies, environmental disturbances and the fact that often actuators used in flight control can exhibit various anomalies such as loss of effectiveness, saturation, or failure, the last of which is our focus in this paper. The nonlinear flight model in the presence of such actuator anomalies can be expressed as:

$$\dot{X} = F\left(X,\ \lambda U\right) \tag{1}$$

where $X \in \Re^n$ is the state vector that can be measured, and U is a scalar input. $\lambda \in \Re$ is a nonzero unknown constant, which represents loss of effectiveness and/or reversal in control input, with $\lambda = 1$ representing the case with no uncertainty. The goal is therefore to determine the input U so that the state X tracks the desired command signals in the presence of uncertainties in λ. Since λ is unknown, equilibrium points and trim inputs in (1) may be unknown. Hence, (1) is linearized at $(X_0,\ U_0)$, such that $F(X_0, U_0) = 0$, which corresponds to the known trim point in the absence of any uncertainties. This leads to the linearized dynamics as

$$\dot{x}_p = A_{p\lambda}x_p + b_{p\lambda}u + d_p + h_p(x_p, u) \tag{2}$$

where

$$A_{p\lambda} = \left.\frac{\partial F(X,\lambda U)}{\partial X}\right|_{(X_0,U_0)},\quad b_{p\lambda} = \left.\frac{\partial F(X,\lambda U)}{\partial U}\right|_{(X_0,U_0)},$$
$$d_p = F(X_0, \lambda U_0),$$

are unknown and $h_p(x_p, u)$ is an unknown nonlinearity that represents higher order terms of x_p and u. In this paper, we assume that h_p depends only on x_p, $F(X, U)$ is affine in U so that $b_{p\lambda} = b_p\lambda$ and $d_p = b_p\lambda d$, and $(A_{p\lambda}, \lambda b_p)$ is controllable. Consequently, (2) is rewritten as

$$\dot{x}_p = A_{p\lambda}x_p + b_p\lambda(u + d + h(x_p)). \tag{3}$$

The nonlinearity can be expressed as

$$h(x_p) = K_0^\top \Phi(x_p) + \epsilon(x_p) \tag{4}$$

where $K_0 \in \Re^N$ is a vector of unknown parameters and $\Phi(x_p) \in \Re^N$ is an approximator of local support functions. $\epsilon(x_p)$ is an approximation error.

The focus of this paper is to evaluate the stability of (3) in the presence of time-delays, unmodeled dynamics, and nonlinearity. That is, we consider a linear plant of the form

$$\dot{x}_p = A_{p\lambda}x_p + b_p\lambda(v + d + h(x_p)), \quad v = G_p(s, \tau)u \tag{5}$$

where the perturbations due to (a) time-delays, (b) unmodeled dynamics, and (c) nonlinearities are

(a) $G_p(s, \tau) = e^{-\tau s}$ and $h(x_p) = 0$,
(b) $G_p(s, \tau) = p_\Delta(s)/q_\Delta(s)$ and $h(x_p) = 0$, and
(c) $G_p(s, \tau) = 1$ and $h(x_p) \neq 0$.

While our focus for the most part is on cases (a) and (c), all results derived here are applicable to case (b) as well.

We proceed to address case (a) by considering Pade approximations of the function $e^{-\tau s}$. In particular, we use the $(1,1)$ and $(2,2)$ Pade approximations of $e^{-\tau s}$ (2) which correspond to a first-order and a second-order rational approximations, respectively. In this paper, we focus on the $(1,1)$ and $(2,2)$ Pade approximations,

$$\begin{aligned} e^{-\tau s} &\cong \frac{-s + 2/\tau}{s + 2/\tau}, \\ e^{-\tau s} &\cong \frac{s^2 - (6/\tau)s + 12/\tau^2}{s^2 + (6/\tau)s + 12/\tau^2} \end{aligned} \tag{6}$$

Using these approximations, the plant in (5) can be written as

$$\dot{x}_p = A_{p\lambda}x_p + b_p\lambda(u + \eta + d + h(x_p)), \; \eta = \Delta_i(s)u \tag{7}$$

where

$$\Delta_1(s) = \frac{-2s}{s + 2/\tau}, \; \Delta_2(s) = \frac{-(12/\tau)s}{s^2 + (6/\tau)s + 12/\tau^2}. \tag{8}$$

The problem considered in this paper is the design of an adaptive controller for the plant in (7) such that stability is guaranteed and the best possible tracking performance is obtained in the presence of the perturbations η and $h(x_p)$.

3 Adaptive Controller

3.1 Reference Model

In order to ensure that all available information about the plant is utilized to obtain the best performance, the adaptive controller is designed in augmentation with a nominal controller. The nominal controller input, u_{nom}, is chosen as

$$u_{nom} = k_x^\top x + k_r r \tag{9}$$

where $k_x \in \Re^n$ and $k_r \in \Re$ are the nominal feedback and feed-forward gain, which are chosen so that the nominal controller input optimizes the performance in the absence of uncertainties and time-delay. r is a command signal with $|r| \leq r_0$ and $|\dot{r}| \leq r_1$ where r_0 and r_1 represent magnitude and rate limits respectively. Therefore, the reference model that the plant in (7) is required to track is generated by combining the nominal controller with (9) as

$$\dot{x}_m = A_m x_m + b_m r \tag{10}$$

where

$$A_m = A_p + b_p k_x^\top, \quad b_m = b_p k_r.$$

Since $(A_p,\, b_p)$ is controllable, k_x can be chosen to ensure A_m is Hurwitz.

3.2 Adaptive Controller Design

An adaptive controller is designed based on the modeled part of the plant in (3) obtained by neglecting $\Delta_1(s)$, $\Delta_2(s)$, and the nonlinearity $h(\cdot)$. The overall control input, u, is designed as

$$u = \theta^\top \omega = \begin{bmatrix} \theta_x^\top & \theta_{nn}^\top & \theta_r & \theta_d \end{bmatrix} \begin{bmatrix} x_p \\ \beta \Phi(x_p) \\ r \\ 1 \end{bmatrix}. \tag{11}$$

and $\theta \in \Re^{n+N+2}$ is the adaptive controller parameter and $\beta = 0$ in cases (a) and (b) while $\beta = 1$ in case (c). The adaptive law is designed as in (3)

$$\dot{\theta} = -\Gamma \omega e^\top P b_p sign(\lambda) - \sigma \theta \tag{12}$$

where $P = P^\top > 0$ is the solution of $A_m^\top P + PA_m = -Q$ for a given $Q = Q^\top > 0$, and $\Gamma = \Gamma^\top > 0$ is a diagonal matrix which represents adaptation rate. Since the adaptive controller is designed based on the modeled part of the plant, the same adaptive law is utilized regardless of the Pade approximations. We also note that $\theta_{nn}(0)$ is set to zero in cases (a) and (b).

3.3 Stability Analysis

Assumption 1. *There exist ideal gains, $\theta_x^* \in \Re^n$, $\theta_{nn}^* \in \Re^N$, $\theta_r^* \in \Re$, and $\theta_d^* \in \Re$ such that*

$$\begin{aligned} A_p + b_p\lambda\theta_x^{*\top} &= A_m, \quad -\theta_{nn}^{*\top} = K_0^\top, \\ b_p\lambda\theta_r^* &= b_m, \quad \theta_d^* = -d. \end{aligned} \tag{13}$$

We define the tracking error to be $e = x - x_m$ and the adaptive parameter error to be $\tilde{\theta} = \theta - \theta^*$. Subtracting (10) from (7), the error equations can be obtained as

$$\dot{e} = A_m e + b_p\lambda\left(\tilde{\theta}^\top\omega + \eta + \epsilon(x_p)\right). \tag{14}$$

In the absence of unmodeled dynamics and nonlinearities, i.e. $\eta = 0$ and $h(x_p) = 0$, it is straight forward to establish the global boundedness and asymptotic tracking of the overall adaptive system in (14) and (8) using standard Lyapunov-based analysis and robustness arguments in (3; 6).

4 Delay Margins

We now derive stability properties of the adaptive system in (7) in the presence of the perturbations, η. This corresponds to case (b) specified in Sect. 2. We first introduce the concept of semi-global boundedness.

Definition 1. *A system*

$$\dot{x} = G(x,\ u) \tag{15}$$

is said to be semi-globally bounded if, for each compact subset $E \subset \Re^n$ where $E^* \subset E$ for a compact subset $E^* \subset \Re^n$, there exists a feedback control input $u = u(x)$ such that the solutions $x(t)$ of the corresponding closed loop system

$$\dot{x} = G(x,\ u(x)) \tag{16}$$

remain inside E (4).

It should be noted that when there is no disturbance, i.e. equilibrium points exist, E^* shrinks to those equilibrium points and semi-global boundedness can be conversed to semi-global stability as in (4). In what follows, we derive the delay margins for the case of (1,1) and (2,2) Pade approximations respectively.

4.1 (1,1) Pade Approximation ($\eta = \Delta_1(s)u$)

To analyze the stability of the closed loop system in the presence of the unmodeled dynamics $\Delta_1(s)$, we express the unmodeled dynamics as the state space form of

$$\dot{\eta} = -(2/\tau)\eta - 2\dot{u}. \tag{17}$$

Definition 2. *$y \in \Re^{2n+2}$ is defined as $y = [e^\top \ \tilde{\theta}^\top \ \eta]^\top$.*

Theorem 1. *For the closed-loop system in (7) with the controller in (11) in the presence of perturbation due to (1,1) Pade approximation, there exist $\tau_{m_1} > 0$ and $y_{\max} > 0$ such that for $0 < \tau < \tau_{m_1}$ and $\|y(t_0)\| < y_{\max}$, $x(t)$, $\theta(t)$, and $\eta(t)$ have semi-globally bounded trajectories with respect to τ for all $t \geq t_0$.*

Proof. Assuming that $\Gamma = \gamma I$ with $\gamma > 0$, we propose a Lyapunov candidate function as

$$V = e^\top Pe + \frac{|\lambda|}{\gamma}\tilde{\theta}^\top\tilde{\theta} + \frac{\tau}{2}(e + \lambda b_p\eta)^\top P(e + \lambda b_p\eta). \tag{18}$$

The time-derivative of V can be obtained along the trajectories of (12), (14), and (16) as

$$\begin{aligned}\dot{V} &= -e^\top Qe - \frac{2\sigma|\lambda|}{\gamma}\tilde{\theta}^\top\tilde{\theta} - 2\lambda^2 b_p^\top Pb_p\eta^2 \\ &+ \tau\Big[-\frac{1}{2}e^\top Qe + \lambda b_p^\top Pe\tilde{\theta}^\top\omega + \lambda b_p^\top Pe\eta - 2\lambda b_p^\top Pe\dot{u} \\ &+ \lambda b_p^\top PA_m e\eta + \lambda^2 b_p^\top Pb_p\tilde{\theta}^\top\omega\eta + \lambda^2 b_p^\top Pb_p\eta^2 \\ &- 2\lambda^2 b_p^\top Pb_p\eta\dot{u}\Big] - \frac{2\sigma|\lambda|}{\gamma}\tilde{\theta}^\top\theta^*.\end{aligned} \tag{19}$$

For simplicity, we define A as

$$\begin{aligned}A = &-\frac{1}{2}e^\top Qe + \lambda b_p^\top Pe\tilde{\theta}^\top\omega + \lambda b_p^\top Pe\eta - 2\lambda b_p^\top Pe\dot{u} \\ &+ \lambda b_p^\top PA_m e\eta + \lambda^2 b_p^\top Pb_p\tilde{\theta}^\top\omega\eta \\ &+ \lambda^2 b_p^\top Pb_p\eta^2 - 2\lambda^2 b_p^\top Pb_p\eta\dot{u}.\end{aligned} \tag{20}$$

From (11), we have $\dot{u} = \dot{\theta}^\top \omega + \theta^\top \dot{\omega}$. In order to find an upper bound of A in terms of $\|e\|$, $\|\tilde{\theta}\|$, and $|\eta|$, we need to find an upper bound of $-2\lambda b_p^\top Pe\dot{u}$ and $-2\lambda^2 b_p^\top Pb_p\dot{u}\eta$. By substituting $\dot{u} = \dot{\theta}^\top \omega + \theta^\top \omega$ into $-2\lambda b_p^\top Pe\dot{u}$, we have

$$\begin{aligned} -2\lambda b_p^\top Pe\dot{u} =& 2|\lambda|\gamma e^\top Pb_p b_p^\top Pe\omega^\top \omega + 2\lambda\sigma b_p^\top Pe\theta^\top \omega \\ &- 2\lambda b_p^\top Pe\theta^\top \dot{\omega}. \end{aligned} \tag{21}$$

We note that $\|\cdot\|$ denotes the vector 2-norm and $\|\cdot\|_{ind}$ is the matrix norm induced by the vector 2-norm. From the bounds on r and $\dot{r}$, we have

$$\|x_m\| \le kr_0, \quad \|\dot{x}_m\| \le (\|A_m\|_{ind} k + \|b_m\|)\, r_0 \tag{22}$$

for a constant $k > 0$. Following inequalities can be useful to find an upper bound of (21):

$$\begin{aligned} \omega^\top \omega &\le \|e\|^2 + 2kr_0\|e\| + (k^2+1)r_0^2 + d_0^2, \\ \theta^\top \omega &\le \|\tilde{\theta} + \theta^*\| \left(\|x_p\| + |r| + |d|\right) \\ &\le \|e\|\|\tilde{\theta}\| + \|\theta^*\|\|e\| + [(k+1)r_0 + d_0]\|\tilde{\theta}\| \\ &+ [(k+1)r_0 + d_0]\|\theta^*\|. \end{aligned} \tag{23}$$

where $|d| \le d_0$. From (23), we also have

$$\begin{aligned} \theta^\top \dot{\omega} \le& \left(\|\tilde{\theta}\| + \|\theta^*\|\right) \Big\{ \|A_m\|_{ind}\|e\|^2 \\ &+ |\lambda|\|b_p\|\|e\| \left[\|e\|\|\tilde{\theta}\| + (k+1)r_0\|\tilde{\theta}\| + |\eta|\right] \\ &+ 2\left[(\|A_m\|_{ind} k + \|b_m\|)\, r_0 + r_1\right]\|e\| \Big\} \end{aligned} \tag{24}$$

Combining (23) and (24), we have an upper bound of A in descending powers of $\|e\|$, $\|\tilde{\theta}\|$, and $|\eta|$. We now have an bound on A as

$$A \le A_4 + A_3 + A_2 + A_1. \tag{25}$$

where A_i denotes the summation of i powers of $\|e\|$, $\|\tilde{\theta}\|$, and $|\eta|$ as

$$A_i = \sum_{\substack{p+q+s=i \\ p,q,s \in \mathbb{N}\cup\{0\}}} C_{p,q,s} \|e\|^p \|\tilde{\theta}\|^q |\eta|^s \tag{26}$$

and $C_{p,q,s}$ is the coefficient corresponding to $\|e\|^p\|\tilde{\theta}\|^q|\eta|^s$ and due to the limited space, $C_{p,q,s}$'s are in Appendix. In order to find an upper bound on A_i, we define constants a_i for $1 \le i \le 4$ such that

$$
\begin{aligned}
a_4 &= \max\left\{C_{4,0,0},\ \frac{C_{3,0,1}}{4},\ \frac{C_{2,2,0}}{6},\ \frac{C_{1,2,1}}{12}\right\} \\
a_3 &= \max\Big\{C_{3,0,0},\ \frac{C_{2,1,0}}{3},\ \frac{C_{2,0,1}}{3},\ \frac{C_{1,2,0}}{3},\ \frac{C_{0,2,1}}{3}, \\
&\qquad \frac{C_{0,1,2}}{3},\ \frac{C_{1,1,1}}{6}\Big\} \\
a_2 &= \max\left\{C_{2,0,0},\ C_{0,0,2},\ \frac{C_{1,1,0}}{2},\ \frac{C_{1,0,1}}{2},\ \frac{C_{0,1,1}}{2}\right\} \\
a_1 &= \max\left\{C_{1,0,0},\ C_{0,0,1}\right\}.
\end{aligned}
\tag{27}
$$

From the definition of y, we obtain

$$
(\|e\| + \|\tilde{\theta}\| + |\eta|)^2 \leq 2\left(\|e\|^2 + \|\tilde{\theta}\|^2 + |\eta|^2|\right) = 2\|y\|^2. \tag{28}
$$

Using the inequality in (28) and constants defined in (27), we have

$$
\begin{aligned}
& A_4 + A_3 + A_2 + A_1 \\
& \leq 4a_4\|y\|^4 + 2\sqrt{2}a_3\|y\|^3 + 2a_2\|y\|^2 + \sqrt{2}a_1\|y\|.
\end{aligned}
\tag{29}
$$

We can rewrite (19) as

$$
\begin{aligned}
\dot{V} \leq & - b_2\|y\|^2 + \tau(4a_4\|y\|^4 + 2\sqrt{2}a_3\|y\|^3 + 2a_2\|y\|^2 \\
& + \sqrt{2}a_1\|y\|) + b_1\|y\|
\end{aligned}
\tag{30}
$$

where

$$
b_2 = \min\left\{q_0,\ \frac{2\sigma|\lambda|}{\gamma}, 2\lambda^2 b_p^\top P b_p\right\},\ b_1 = \frac{2\sigma|\lambda|}{\gamma}\|\theta^*\|. \tag{31}
$$

We can rewrite the inequality in (30) as

$$
\begin{aligned}
\dot{V} \leq & 4a_4\tau\|y\|\left[\|y\|^3 + \frac{a_3}{\sqrt{2}a_4}\|y\|^2 + \frac{1}{2a_4}\left(a_2 - \frac{b_2}{2\tau}\right)\|y\|\right. \\
& \left. + \frac{1}{2\sqrt{2}a_4}\left(a_1 + \frac{b_1}{\sqrt{2}\tau}\right)\right].
\end{aligned}
\tag{32}
$$

To accommodate the analysis simply, we substitute $\|y\|$ with $z > 0$ and set $f(z)$ as

$$
f(z) = z^3 + az^2 + bz + c \tag{33}
$$

where

$$
\begin{aligned}
a &= \frac{a_3}{\sqrt{2}a_4}, \quad b = \frac{1}{2a_4}\left(a_2 - \frac{b_2}{2\tau}\right), \\
c &= \frac{1}{2\sqrt{2}a_4}\left(a_1 + \frac{b_1}{\sqrt{2}\tau}\right).
\end{aligned}
\tag{34}
$$

Since $f(z)$ is a third-order polynomial, $f(z)$ is continuous. From the facts that $a > 0$ and $c > 0$, $f(0) > 0$ and $\lim_{z \to +\infty} f(z) = +\infty$. From these facts, following two cases are possible which are mutually exclusive:

(a)$f(z)$ does not intersect with z-axis at $[0, \infty)$.
(b)$f(z)$ intersects with z-axis twice at $[0, \infty)$.

To show that case (a) cannot occur for $\tau \in (0,\ \tau_{m_1})$, it suffices to prove that (i) $z_1 = \arg\min_{z \in [0,\infty)} f(z)$ exists and (ii) $f(z_1) < 0$. We prove (i) and (ii) as follows. Roots of $f'(z) = 0$ are

$$z_1 = \frac{-a + \sqrt{a^2 - 3b}}{3}, \qquad z_2 = \frac{-a - \sqrt{a^2 - 3b}}{3}. \tag{35}$$

In order for z_1 and z_2 to exist, $a^2 - 3b > 0$. That is,

$$\frac{1}{\tau} > \frac{2a_2}{b_2} - \frac{2a_3^2}{3a_4 b_2}. \tag{36}$$

In order for $z_1 > 0$, we also requires $b > 0$, i.e.

$$\frac{1}{\tau} > \frac{2a_2}{b_2}. \tag{37}$$

When we choose τ_{m_1} such that $\tau_{m_1} < b_2/2a_2$, the proof of (i) is completed. Using $3z_1^2 + 2az_1^2 + b = 0$, the minimum of $f(z)$ can be simplified as

$$\begin{aligned} \min_{z \in [0,\infty)} f(z) &= f(z_1) = z_1^3 + az_1^2 + bz_1 + c \\ &= -\frac{2}{27}\left(\sqrt{a^2 - 3b}\right)^3 + \frac{2}{27}a^3 - \frac{1}{3}ab + c. \end{aligned} \tag{38}$$

Since b and c are a function of τ, $f(z_1)$ is a function of τ which is denoted by $g(\tau)$ and given by

$$\begin{aligned} g(\tau) = &-\frac{2}{27}\left[\frac{a_3^2}{2a_4^2} - \frac{3}{2a_4}\left(a_2 - \frac{b_2}{2\tau}\right)\right]^{\frac{3}{2}} + \frac{2a_3^3}{54\sqrt{2}a_4^3} \\ &- \frac{a_3}{6\sqrt{2}a_4^2}\left(a_2 - \frac{b_2}{2\tau}\right) + \frac{1}{2\sqrt{2}a_4}\left(a_1 + \frac{b_1}{\sqrt{2}\tau}\right). \end{aligned} \tag{39}$$

$g(\tau)$ is a continuous function of τ at $(0,\ \infty)$ and, $\lim_{\tau \to 0} g(\tau) = -\infty$ and $g(b_2/2a_2) > 0$ so that by the intermediate value theorem, there exists $\tau_{m_1} \in (0,\ b_2/2a_2)$ such that

$$g(\tau_{m_1}) = 0 \quad \text{and} \quad g(\tau) < 0, \quad \forall \tau \in (0,\ \tau_{m_1}). \tag{40}$$

This proves that $f(z_1) < 0$ for $\tau \in (0, \tau_{m_1})$, which establishes (ii).

Because $f(z_1) < 0$ for $\tau \in (0,\ \tau_{m_1})$, there exist $y_{\min}$ and $y_{\max}$ such that $0 < y_{\min} < z_1 < y_{\max}$ and $f(y_{\min}) = f(y_{\max}) = 0$ by the intermediate value theorem. Defining a set $\mathbf{A}$ as

$$\mathbf{A} = \{y | y_{\min} < \|y\| < y_{\max}\}, \tag{41}$$

it leads that $\dot{V} < 0$ in $\mathbf{A}$. As $\tau \to 0$, $z_1 \to \infty$ and $y_{\max} \to \infty$. In this sense, the stability result obtained is semi-global as $\mathbf{A}$ expands to the whole space as $\tau \to 0$.

Remark 1. *For the (1,1) Pade approximation, τ_{m_1} is the guaranteed time-delay margin.*

4.2 (2, 2) Pade Approximation ($\eta = \Delta_2(s)u$)

The output due to the unmodeled dynamics $\Delta_2(s)$ is expressed as

$$\begin{aligned} \dot{\xi} &= \begin{bmatrix} 0 & 1 \\ -12/\tau^2 & -6/\tau \end{bmatrix} \xi + \begin{bmatrix} 0 \\ -12/\tau \end{bmatrix} \dot{u}, \\ \eta &= [1\ 0]\,\xi. \end{aligned} \tag{42}$$

Theorem 2. *For the closed-loop system in* (7) *with the controller in* (11) *in the presence of perturbation due to (2,2) Pade approximation, there exist $\tau_{m_2} > 0$ and $y_{\max} > 0$ such that for $0 < \tau < \tau_{m_2}$ and $\|y(t_0)\| < y_{\max}$, $x(t)$, $\theta(t)$, and $\xi(t)$ have semi-globally bounded trajectories with respect to τ for all $t \geq t_0$.*

Proof. A Lyapunov candidate function is chosen as

$$\begin{aligned} W &= e^\top P e + \frac{|\lambda|}{\gamma}\tilde{\theta}^\top \tilde{\theta} \\ &\quad + \tau \left(e + \frac{b_p \lambda c^\top \xi}{2}\right)^\top P \left(e + \frac{b_p \lambda c^\top \xi}{2}\right) \\ &\quad + \tau \frac{\lambda^2 b_p^\top P b_p}{6} \eta^2 \end{aligned} \tag{43}$$

where $c^\top = [1/\tau \quad 1/6]$. The proof proceeds in an identical manner to that of Theorem 1.

Remark 2. *For the (2,2) Pade approximation, τ_{m_2} is the guaranteed time-delay margin.*

5 Nonlinearity Margins

In this section, we discuss stability properties of the adaptive system in the presence of the nonlinearity $h(x_p)$ which corresponds to case (c) in Sect. 2. The underlying error equations are given by (14) with $\eta = 0$, and the adaptive laws are given by (12), with $\beta = 1$. The local support functions $\Phi(x_p)$ defined in (11) are placed around x_{p_i} so that the approximation error $\epsilon(x_p)$ is bounded around x_{p_i} and that it is bounded by a function of x_p otherwise. This is stated in the following assumption.

Assumption 2. *The approximation error, $\epsilon(x_p)$, is bounded as*

$$\|\epsilon(x_p)\| \leq \mu\left(\|x_p\|^q + \mu_c\right). \tag{44}$$

where μ and μ_c are positive constants and $q > 0$. $\|\cdot\|$ denotes vector 2-norm.

In what follows, we will show that an adaptive controller with a neural network can be determined such that semi-global boundedness of the closed loop system is guaranteed under Assumptions 1 and 2 with respect to μ. For efficiency of notation, $p_{\max}$ is defined by using the induced norm such that the property is given by

$$\|e^\top P b_p\| \leq p_{\max}\|e\| \tag{45}$$

and following constants are defined as

$$\begin{aligned} q_{\min} &= \min(eig(Q)), \\ \gamma_{\max} &= \max(eig(\Gamma)), \ \gamma_{\min} = \min(eig(\Gamma)). \end{aligned} \tag{46}$$

Theorem 3. *Under Assumptions 1 and 2, there exists a $\mu^* > 0$ such that for $0 < \mu \leq \mu^*$, the closed-loop system in (7) with the controller in (11) has a bounded solution of $(x_p(t),\ \theta(t))$ for all $t \geq t_0$ if*

$$\|e(t_0)\| \leq \sqrt[q]{\frac{cq_{\min}}{2|\lambda|p_{\max}} \cdot \frac{1}{\mu} - \mu_c - x_0} \tag{47}$$

for some constant $c > 0$. Furthermore, the bound on $\|e(t_0)\|$ tends to ∞ as $\mu \to 0$. Hence $(x_p(t),\ \theta(t))$ is semi-globally bounded with respect to μ.

Proof. We propose a Lyapunov candidate function as

$$V(e,\tilde{\theta}) = e^\top Pe + |\lambda|\tilde{\theta}^\top \Gamma^{-1}\tilde{\theta}. \tag{48}$$

The time derivative of V along the trajectories of (12) and (14) is

$$\dot{V} = -e^\top Qe + 2e^\top Pb_p\lambda\epsilon(x_p) - \sigma|\lambda|\tilde{\theta}^\top \Gamma^{-1}\theta \tag{49}$$

From Assumption 1, we have

$$\begin{aligned}\dot{V} \leq & -q_{\min}\|e\|\left[\|e\| - 2\mu\frac{|\lambda|p_{\max}}{q_{\min}}\left(\|x_p\|^q + \mu_c\right)\right] \\ & -\sigma\frac{|\lambda|}{\gamma_{\max}}\|\tilde{\theta}\|^2 + \sigma\frac{|\lambda|}{\gamma_{\min}}\|\tilde{\theta}\|\end{aligned} \tag{50}$$

We suppose that

$$2\mu\frac{|\lambda|p_{\max}}{q_{\min}}\left(\|x_p\|^q + \mu_c\right) \leq c \tag{51}$$

for a constant $c > 0$. Then, we have

$$\begin{aligned}\dot{V} \leq & -q_{\min}\left(\|e\| - \frac{c}{2}\right)^2 - \sigma\frac{|\lambda|}{\gamma_{\max}}\left(\|\tilde{\theta}\| - \frac{\gamma_{\max}}{2\gamma_{\min}}\right)^2 \\ & + q_{\min}\left(\frac{c}{2}\right)^2 + \sigma\frac{|\lambda|\gamma_{\max}}{4\gamma_{\min}^2}.\end{aligned} \tag{52}$$

We define a compact set as

$$\begin{aligned}\mathbf{D} = \Bigg\{(e,\tilde{\theta})\,\Bigg|\, & q_{\min}\left(\|e\| - \frac{c}{2}\right)^2 + \sigma\frac{|\lambda|}{\gamma_{\max}}\left(\|\tilde{\theta}\| - \frac{\gamma_{\max}}{2\gamma_{\min}}\right)^2 \\ & \leq q_{\min}\left(\frac{c}{2}\right)^2 + \sigma\frac{|\lambda|\gamma_{\max}}{4\gamma_{\min}^2}\Bigg\}.\end{aligned} \tag{53}$$

This shows that if the inequality in (51) holds, $\dot{V} < 0$ outside of $\mathbf{D}$ which guarantees the boundedness of $(e,\ \tilde{\theta})$. In what follows, we will show that there exists a set $\mathbf{M}(e, \tilde{\theta})$ such that (51) holds for all $e \in \mathbf{M}$ and $\mathbf{D} \subset \mathbf{M}$. From the definition of $\mathbf{D}$, we have

$$\|e\| \leq \frac{c}{2} + \sqrt{\left(\frac{c}{2}\right)^2 + \sigma\frac{|\lambda|\gamma_{\max}}{4q_{\min}\gamma_{\min}^2}}, \qquad \forall e \in \mathbf{D}. \tag{54}$$

The inequality in (51) can be rewritten as

$$\|x_p\|^q \leq \frac{cq_{\min}}{2|\lambda|p_{\max}} \cdot \frac{1}{\mu} - \mu_c. \tag{55}$$

Defining a set $\mathbf{M}$ as

$$\mathbf{M} = \left\{(e,\ \tilde{\theta})\ \Bigg|\ \|e\| \leq \sqrt[q]{\frac{cq_{\min}}{2|\lambda|p_{\max}} \cdot \frac{1}{\mu} - \mu_c} - x_0\right\}, \tag{56}$$

where $\|x_m(t)\| \leq x_0$. It follows that (55) holds for all $(e,\ \tilde{\theta}) \in \mathbf{M}$. In order to ensure $\mathbf{D} \subset \mathbf{M}$, we need the following condition as

$$\frac{c}{2}+\sqrt{\left(\frac{c}{2}\right)^2+\sigma\frac{|\lambda|\gamma_{\max}}{4q_{\min}\gamma_{\min}^2}} \leq \sqrt[q]{\frac{cq_{\min}}{2|\lambda|p_{\max}}\cdot\frac{1}{\mu}-\mu_c}-x_0. \tag{57}$$

It follows that (57) is satisfied for all $\mu \in (0,\ \mu^*]$ if we choose μ^* as

$$\mu^* = \frac{cq_{\min}/2|\lambda|p_{\max}}{\left[\left(x_0+\frac{c}{2}+\sqrt{\left(\frac{c}{2}\right)^2+\sigma\frac{|\lambda|\gamma_{\max}}{4q_{\min}\gamma_{\min}^2}}\right)^q+\mu_c\right]}. \tag{58}$$

Therefore, we have $\mathbf{D} \subset \mathbf{M}$ for all $0 < \mu \leq \mu^*$. This implies that for $(e(t_0), \tilde{\theta}(t_0)) \in \mathbf{M} - \mathbf{D}$, $(e(t), \tilde{\theta}(t))$ has a bounded solution for all $t \geq t_0$. This in turn concludes that $(x_p(t), \theta(t))$ in the closed-loop system with the proposed adaptive controller has a bounded solution for all $t \geq t_0$. The complete the boundedness of the closed-loop system. Furthermore, as μ tends to zero,

$$\sqrt[q]{\frac{cq_{\min}}{2|\lambda|p_{\max}}\cdot\frac{1}{\mu}-\mu_c}-x_0 \to \infty \tag{59}$$

which implies that $\mathbf{M}$ tends to the whole space of $(e, \tilde{\theta})$. In this sense, the stability result becomes semi-global with respect to μ.

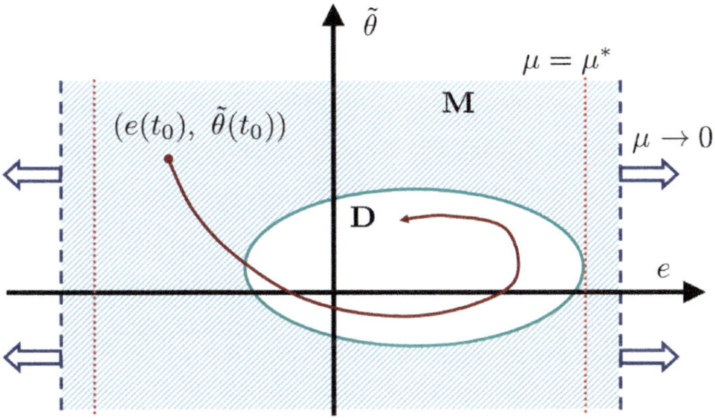

Fig. 1 Geometrical interpretation of the semi-global stability

The stability result is interpreted geometrically in Fig. 1. When $\mu = \mu^*$, the compact set $\mathbf{D}$ is critically contained in $\mathbf{M}$. For $\mu < \mu*$, there exists the region, $\mathbf{M} - \mathbf{D}$, where the boundedness of $(e(t),\ \tilde{\theta}(t))$ is guaranteed with $(e(t_0),\ \tilde{\theta}(t_0)) \in \mathbf{M} - \mathbf{D}$. As μ tends to zero, $\mathbf{M}$ extends to the wholes space as shown in Fig. 1 so that the stability result becomes global.

5.1 Interpretation of Theorem 3

Theorem 3 establishes that stability of the closed system is guaranteed when the initial condition, $e(t_0)$, satisfies (51). That is, the size of initial conditions that guarantees boundedness of the closed-loop system is dependent on μ. As μ decreases, this size expands to include the entire state-space. It should also be noted that Assumption 2 shows that a smaller μ implies that the approximation error of the residual nonlinearity using the approximator of local support functions needs to be smaller. This in turn can be accomplished by increasing N, the number of the support functions for a given μ and μ_c. For a given nonlinearity with a fixed number of local support functions, namely N, the approximation error is fixed which in turn determines μ in (44). This implies that in such a case, there is a certain compact domain M of $e(t_0)$ for which boundedness holds; no guaranties of boundedness can be given if $e(t_0)$ lies outside **M**.

In the following section we explore the relationship between $e(t_0)$, N, and μ in the context of a flight control example.

5.2 Numerical Model: Hypersonic Vehicle

The short period dynamics of NASA X-15 is employed to validate the theoretical findings from the previous section. The short period dynamics are given by

$$\dot{x}_p = A_p x_p + b_p \lambda \left[u + d + h(x_p) \right] \tag{60}$$

where $x_p = [\alpha \ q]^\top$ in *deg* and *deg/s* respectively, and u is the elevator input in *deg*. The matrices A_p and b_p correspond to linearized dynamics obtained by trimming the overall nonlinear dynamics at 60,000 ft with the speed of 1,930 ft/s and are given by

$$A_p = \begin{bmatrix} -0.2950 & 1.0000 \\ -13.0798 & -0.2084 \end{bmatrix}, \ b_p = \begin{bmatrix} 0 \\ -9.4725 \end{bmatrix}. \tag{61}$$

λ represents the uncertainty in the control input and was set as 0.5 which implies a 50% loss of the elevator input effectiveness. $h(x_p)$ represents a pitch break phenomenon which solely depends on α as

$$h(x_p) = -\alpha^2. \tag{62}$$

In the absence of uncertainty ($\lambda = 0$) and nonlinearity ($h(x_p) = 0$), the nominal controller is designed so as to ensure the closed-loop performance as

$$u_{nom} = k_x^\top x + k_r r \tag{63}$$

where r is a command signal for angle of attack. The feedback gain, k_x, is found by minimizing

$$J = \int_0^\infty x_p^\top \left(\bar{Q} + k_x^\top \bar{R} k_x \right) x_p dt \tag{64}$$

Table 1 Constants to compute μ^*

Symbols	Quantities	Symbols	Quantities
c	1	$\gamma_{\max}$	100
$q_{\min}$	1	$\gamma_{\min}$	10
$p_{\max}$	0.4643	σ	0.1
λ	0.5	x_0	3.47
q	3		

where x_p is subject to (60) with $\lambda = 0$ and $h(x_p) = 0$. The suitable closed-loop response is obtained with $\bar{Q} = diag([2\ 1])$ and $\bar{R} = 1$. The feed-forward gain, k_r, is determined as $k_r = 1/g_\alpha$ where $[g_\alpha\ g_q]^\top = -A_m^{-1} b_p$ based on the assumption that r is a step input. Combining the nominal controller with the plant, we have a reference model as

$$\dot{x}_m = A_m x_m + b_m r \tag{65}$$

where $b_m = b_p k_r$. The matrices A_m and b_m were given by

$$A_m = \begin{bmatrix} -0.295 & 1.000 \\ -50.934 & -10.153 \end{bmatrix},\ b_m = \begin{bmatrix} 0 \\ 53.929 \end{bmatrix}. \tag{66}$$

We simulated the reference model in (65) with unit-step input to obtain $\|x_m(t)\| \leq 3.47$ when $x_m(t_0) = 0$ is assumed without loss of generality. We chose $Q = I_{2\times 2}$ and solved $A_m^\top P + P A_m = -Q$ to obtain

$$P = \begin{bmatrix} 2.4414 & -0.0043 \\ -0.0043 & 0.0488 \end{bmatrix}. \tag{67}$$

The adaptation rate in (12) can be written as

$$\Gamma = \begin{bmatrix} \Gamma_x & 0 & 0 & 0 \\ 0 & \Gamma_{nn} & 0 & 0 \\ 0 & 0 & \Gamma_r & 0 \\ 0 & 0 & 0 & \Gamma_d \end{bmatrix} \tag{68}$$

where Γ_x, Γ_{nn}, Γ_r, and Γ_d were set to $10 I_{2\times 2}$, $10 I_{N\times N}$, 100, and 10 respectively. To prevent adaptive parameters from being constrained excessively, we set $\sigma = 0.1$ which is sufficiently small. For simplicity, c in (51) was set as 1. We set $q = 3$ because it is obvious that Assumption 2 cannot be satisfied with $q \leq 2$ for $h(x_p) = -\alpha^2$. Substituting constants into (58), we have

$$\mu^* = \frac{1}{0.4643\,(4.4823^3 + \mu_c)}. \tag{69}$$

$\Phi(\alpha)$ is designed with a series of radial basis functions and ith component of $\Phi(\alpha)$ is defined by

$$\Phi_i(\alpha) = \frac{1}{\sqrt{2\pi}} \exp\left(-\frac{(\alpha-\alpha_i)^2}{2}\right). \tag{70}$$

We choose $\alpha_1 = 0$ and $\alpha_i = i$ for $i = 1, 2, \cdots, N$. K_0 in (4) was obtained as

$$K_0 = S^{-1} \begin{bmatrix} f(\alpha_1) \\ \vdots \\ f(\alpha_N) \end{bmatrix} \tag{71}$$

where $S \in \Re^{N\times N}$ and $S_{ij} = \Phi_j(\alpha_i)$ such that $K_0^\top \Phi(\alpha_i) = h(\alpha_i)$ for $i = 1, 2, \cdots, N$.

5.3 Relation between $e(t_0)$, $\epsilon(x_p)$, and N

Noting that (69) holds for any μ_c, we choose $\mu_c = 3$, which leads to $\mu^* = 0.0231$. Then, (47) implies that boundedness follows for all $\|e(t_0)\| \le 1.0123$. In order to satisfy Assumption 2, a suitable approximator should be chosen. We observed that for the above μ^* and μ_c, Assumption 2 can be satisfied for $N = 45$. This result is plotted in Fig. 2.

Suppose that stability needs to be guaranteed for a bigger initial condition $e(t_0)$. Then, μ should be decreased to satisfy (47) and this in turn requires

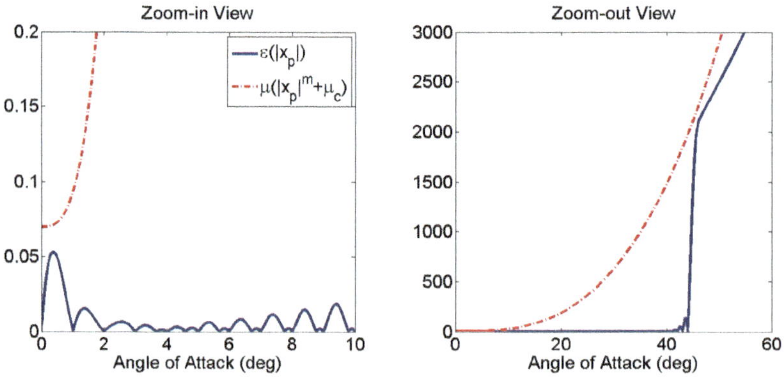

Fig. 2 Approximation error and the upper bound function($N = 45$)

Table 2 Initial conditions and required N

$\|e(t_0)\|$	μ	μ_c	N
1.0123	0.0231	3	45
2.0000	0.0128	5	79
3.0000	0.0077	8	129
4.0000	0.0050	12	201

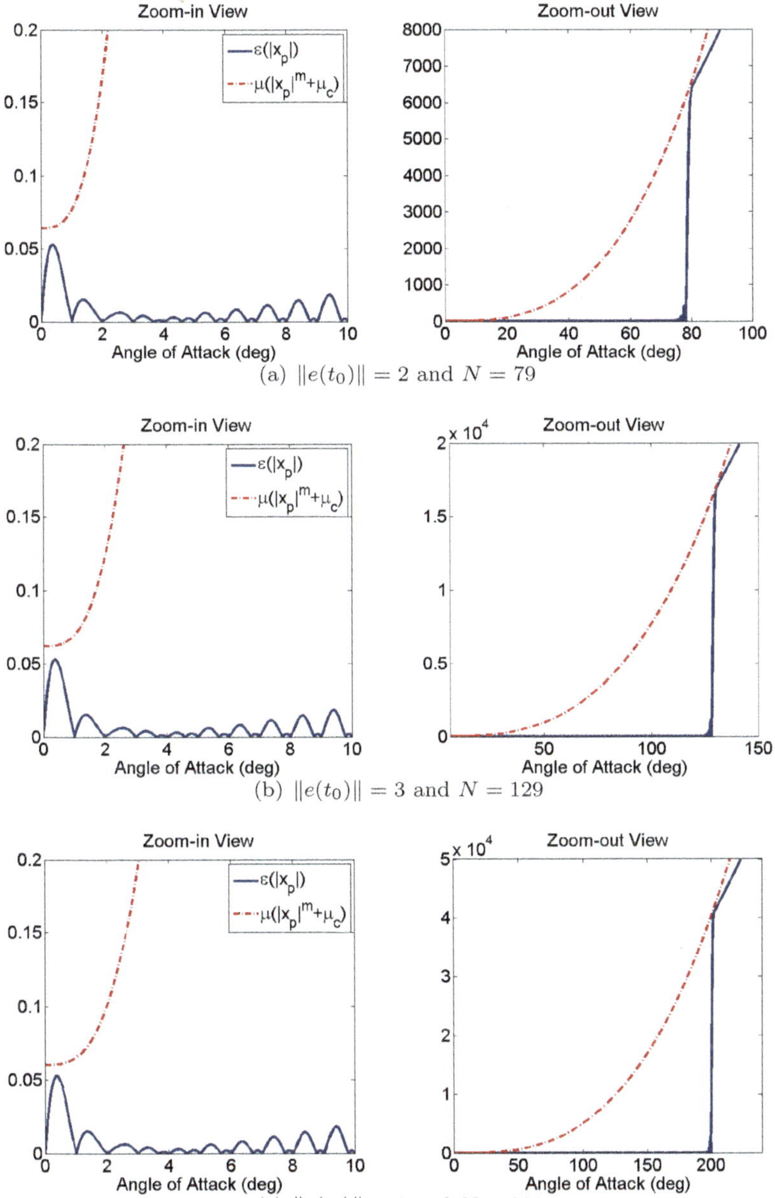

(a) $\|e(t_O)\| = 2$ and $N = 79$

(b) $\|e(t_O)\| = 3$ and $N = 129$

(c) $\|e(t_O)\| = 4$ and $N = 201$

Fig. 3 Approximation error and the upper bound function when (a) $\|e(t_O)\| = 2$, (b) $\|e(t_O)\| = 3$, and (c) $\|e(t_O)\| = 4$

that $\|\epsilon(x_p)\|$ in Assumption 2 should be reduced. Therefore, in order to satisfy Assumption 2, we increase N until Assumption 2 holds. Table 2 shows the relation between the resulting N, μ, and $e(t_0)$. The approximation error and the upper bound based on Assumption 2 were plotted in Fig. 3 respectively. This shows that for every initial condition $x_p(t_0)$, there exists a sufficiently large N for which boundedness holds. It should also be pointed out that the choices for many of the parameters in the above discussion were empirical in nature. More careful investigations that deploy appropriate optimization procedures can lead to a less conservative choice of μ and N. For instance, a much smaller N can be obtained if one were to replace the function $\mu(\|x_p\|^q + \mu_c)$ by an exponential function $\mu_c \exp(\mu\|x_p\|)$. The derivation of boundedness in this case is currently in progress.

References

1. Astrom, K.J., Wittenmark, B.: Adaptive Control, 2nd edn. Addison-Wesley, Reading (1996)
2. Franklin, G.F., Powell, J.D., Emami-Naeini, A.: Feedback Control of Dynamic Systems, 3rd edn. Addison-Wesley, Reading (1994)
3. Ioannou, P.A., Sun, J.: Robust Adaptive Control. Prentice-Hall, Upper Saddle River (1995)
4. Isidori, A.: Nonlinear Control Systems, 3rd edn. Springer, London (1995)
5. Landau, I.D.: Adaptive Control: The Model Reference Approach. Marcel Dekker, New York (1979)
6. Narendra, K.S., Annaswamy, A.M.: Stable Adaptive Systems. Prentice-Hall, Englewood Cliffs (1989)
7. Sastry, S., Bodson, M.: Adaptive Control: Stability, Convergence, and Robustness. Prentice-Hall, Englewood Cliffs (1989)

A Appendix

$C_{p,q,s}$ in (26) are defined as

$$
\begin{aligned}
C_{4,0,0} &= 2\gamma|\lambda|\|b_p^\top P\|^2 \\
C_{3,0,1} &= 2\gamma\lambda^2 b_p^\top P b_p \|b_p^\top P\| \\
C_{2,2,0} &= 2\lambda^2 b_p^\top P b_p \\
C_{1,2,1} &= 2|\lambda|^3 b_p^\top P b_p \|b_p\| \\
C_{3,0,0} &= 4\gamma|\lambda| k r_0 \|b_p^\top P\|^2 \\
C_{2,1,0} &= 2|\lambda|\|b_p^\top P\|\left(\sigma + \|A_m\|_{ind}\right) + 2\lambda^2 b_p^\top P b_p \|\theta^*\| + |\lambda|\|b_p^\top P\| \\
C_{2,0,1} &= 4\gamma\lambda^2 k r_0 b_p^\top P b_p \|b^\top P\| \\
C_{1,2,0} &= 2\lambda^2 b_p^\top P b_p((k+1)r_0 + d_0) \\
C_{1,1,1} &= \lambda^2 b_p^\top P b_p \Big[3 + 2(\sigma + \|A_m\|_{ind} + |\lambda|\|b_p\|\|\theta^*\|)\Big] \\
C_{0,2,1} &= 2|\lambda|^3 b_p^\top P b_p \|b_p\|((k+1)r_0 + d_0) \\
C_{0,1,2} &= 2|\lambda|^3 b_p^\top P b_p \|b_p\| \\
C_{2,0,0} &= 2|\lambda|\|b_p^\top P\|\Big[\gamma\|b_p^\top P\|\left((k^2+1)r_0^2 + d_0^2\right) + (\sigma + \|A_m\|_{ind})\|\theta^*\|\Big] - \tfrac{q_0}{2} \\
C_{1,1,0} &= \Big\{2|\lambda|\|b_p^\top P\|\Big[\sigma((k+1)r_0 + d_0) + (\|A_m\|_{ind}k + \|b_m\|)r_0 + r_1\Big] + \\
&\quad 2\lambda^2 b_p^\top P b_p((k+1)r_0 + d_0)\|\theta^*\|\Big\} \\
&\quad + |\lambda|\|b_p^\top P\|((k+1)r_0 + d_0) \\
C_{1,0,1} &= 2\lambda^2 b_p^\top P b_p \Big[\gamma\|b_p^\top P\|\left((k^2+1)r_0^2 + d_0^2\right) + (\sigma + \|A_m\|_{ind} + 1)\|\theta^*\|\Big] + \\
&\quad |\lambda|\|b_p^\top P\|(\|A_m\|_{ind} + 1) \\
C_{0,1,1} &= 2\lambda^2 b_p^\top P b_p \Big[(k+1)r_0(\sigma + |\lambda|\|b_p\|\|\theta^*\|) + (\|A_m\|_{ind}k + \|b_m\|)r_0 + r_1\Big] \\
&\quad + \lambda^2 b_p^\top P b_p((k+1)r_0 + d_0) \\
C_{0,0,2} &= \lambda^2 b_p^\top P b_p(2|\lambda|\|b_p\|\|\theta^*\| + 1) \\
C_{1,0,0} &= 2|\lambda|\|b_p^\top P\|\|\theta^*\|\Big[\sigma((k+1)r_0 + d_0) + (\|A_m\|_{ind}k + \|b_m\|)r_0 + r_1\Big] \\
C_{0,0,1} &= 2\lambda^2 b_p^\top P b_p \|\theta^*\|\Big[\sigma((k+1)r_0 + d_0) + (\|A_m\|_{ind}k + \|b_m\|)r_0 + r_1\Big].
\end{aligned}
$$

Network Complexity Analysis of Multilayer Feedforward Artificial Neural Networks

Helen Yu

Abstract. Artificial neural networks (NN) have been successfully applied to solve different problems in recent years, especially in the fields of pattern classification, system identification, and adaptive control. Unlike the traditional methods, the neural network based approach does not require a priori knowledge on the model of the unknown system and also has some other significant advantages, such as adaptive learning ability as well as nonlinear mapping ability. In general, the complexity of a neural network structure is measured by the number of free parameters in the network; that is, the number of neurons and the number and strength of connections between neurons (weights). Network complexity analysis plays an important role in the design and implementation of artificial neural networks - not only because the size of a neural network needs to be predetermined before it can be employed for any application, but also because this dimensionality may significantly affect the neural network learning and generalization ability. This chapter gives a general introduction on neural network complexity analysis. Different pruning algorithms for multi-layer feedforward neural networks are studied and computer simulation results are presented.

1 Introduction

There are two fundamental issues in neuro-computation: learning algorithm development and the network topology design. In fact, these two issues are closely related with each other. The learning ability of a neural network is not only a function of time (or training iterations), but also a function of the network structure.

A typical neural network contains an input layer, an output layer, and one or more hidden layers. The number of outputs and the number of inputs

J. Schumann, Y. Liu (Eds.): Appl. of Neural Networks in High Assur. Sys., SCI 268, pp. 41–55.
springerlink.com

are usually fixed (note that in some applications, even the number of inputs can be changed — i.e., there may exist some inputs that are not actually related with the system dynamics and/or solution of the problem); while the number of hidden layers and number of hidden neurons in each hidden layer are parameters that can be specified for each application.

In a feedforward neural network, all the neurons are connected only in forward direction (Fig. 1). It is the class of neural networks that is used most often in system identification and adaptive control (1; 8; 19; 24; 26; 27). As we know from literature, a multi-layer feedforward neural network with correct value of weights (and appropriate transfer functions of the neurons) is capable of approximating any measurable continuous function to any degree of accuracy with only one hidden layer (but infinite number of hidden neurons) (18; 20; 25)]. However, in practice, it is impossible to construct such a neural network with infinite number of hidden nodes. On the other hand, none of the above results indicates how the neural network can be trained to generate the expected output and in fact, any existing learning law cannot guarantee this happening in a finite time period.

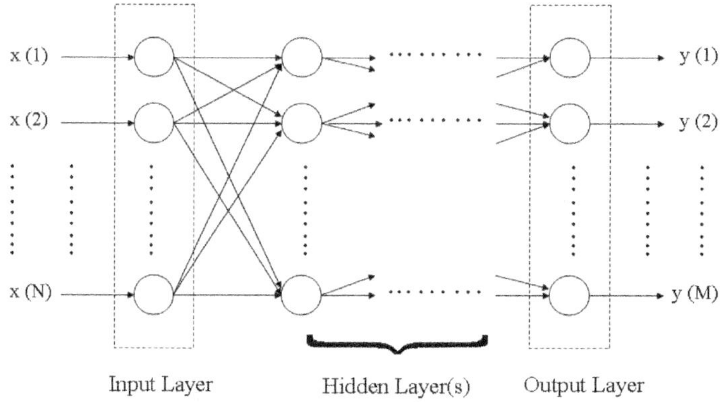

Fig. 1 Multilayer Feed-forward Network

In general, a larger neural network (i.e., more weights and hidden nodes) may yield a faster rate of convergence and is more powerful to solve problems (e.g., to recall patterns). On the other hand, a smaller neural network requires less computation time and is advantageous in real-time environment where speed is crucial. The neural network dimension is also related with the generalization property of a NN (2; 5; 7; 38). For a given set of training pairs, there exist many networks that can learn the underlying mapping function; however, the generalization capability of each network may be different. Generalization indicates the ability of a NN to produce accurate interpolation and/or prediction on data which are not included in the training set. If the structure of a NN is more complicated than necessary, over-fitting may occur. That is, the oversized NN may give exactly the right outputs on training samples, but fails on the other points between and/or beyond samples (Fig. 2). This is similar to the case in curve fitting problem when a high order polynomial is chosen while only a few points are available. Over-fitting is not an issue when a comprehensive training set is available, since all possible input/output pairs are presented and little or no generalization is needed. However, the amount of training data is usually limited; thus a trained network is expected to be able to perform well even on previously unseen data. Therefore, it is very important to find the "optimal", or "appropriate" size of a neural network for a specific application. An ideal neural network should be able to perform well on both training data and unknown testing data while maintaining its form as compact as possible.

An approach to determine the appropriate size of NN using Bayesian model comparison can be found in (40), where the neural networks with different hidden neurons or hidden layers are considered as a set of different models to be evaluated. It is based on the principle that the Bayesian formalism automatically penalizes the models with more complicated structures. A best model is the one which can balance the need of a large likelihood (so that it can fit well with data) and the need of a network with simpler structure.

Some recent research works investigate the optimal neural network problem using evolutionary computation (3; 14; 31; 35). As we know, a genetic algorithm is based on Charles Darwin's theory of natural selection (i.e., "survival of the fittest"). From Darwin's theory, individuals in a population of reproductive organisms inherit traits from their parents; and the desirable traits become more common than the undesirable ones during evolution, due to the fact that the individuals with the desirable traits which "fit" better with the environment are more likely to reproduce.

In a genetic algorithm (39), all the possible solutions are represented by genes which are usually coded as binary strings. As in natural selection process, an initial population of individuals is generated first, with all of their genes randomly selected. Each individual in the population is then sorted based on a certain performance criterion and those individuals with higher fitness levels (according to performance criteria) are more likely to be selected to reproduce the next generation of solutions that may yield a higher fitness index than the parents. The general reproduction operations involve crossover (i.e., exchange genes) and mutation (i.e., randomly change the value of a gene).

A genetic algorithm is a powerful global random search algorithm which finds the optimal solution in parameter space; however, the computation time may be very long.

Other than a lucky guess and/or an extensive search algorithm, there are two fundamental approaches to find the appropriate size of a neural network. The first one is to start with a small network and slowly add more connections to it until an appropriate stopping criterion is satisfied (11; 34). The

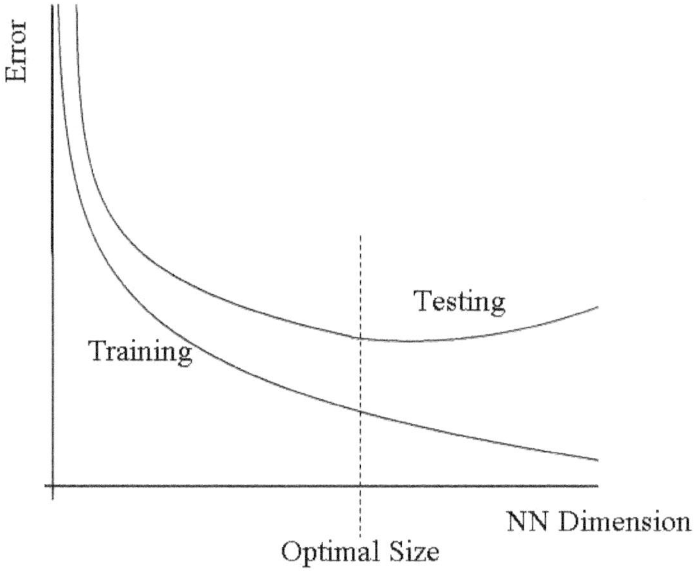

Fig. 2 Relationship between NN dimension, training error, and testing error

difficulties of this approach include when to start the growing process, and where to add the new connections/nodes in the network. In addition, the above procedures may be very tedious and time-consuming. The second approach is to start with a network that is knowingly too large for the data, and then trim it down to the appropriate size. This is called "neural network pruning" and has been studied by many researchers in recent years (12; 15; 16; 17; 29; 30; 32; 33). In this research, we focus on different pruning algorithms for multi-layer feedforward neural networks. Our goal is to reduce the size of a neural network to reduce its computation time while maintaining satisfactory generalization accuracy to improve the overall system performance.

2 Pruning Algorithms

Recent research on pruning algorithms may be classified into two major categories, i.e., the weight decay (or penalty-term) method and the sensitivity based approach. The penalty-term method adds an additional term to the cost function to be minimized so that the unnecessary weights in the network can be eliminated (i.e., driven to zero) during training (28; 36; 37). This penalty term is usually defined as the sum of the quadratic form of weights times a (or several) decay constant(s), as shown in Eq. (1):

$$J_0 = J + \frac{1}{2}\mu \sum_{i,j} w_{ij}^2 \tag{1}$$

where $J_0(\cdot)$ is the overall objective function; w_{ij} is the weight/connection of the neural network from node i to node j; μ is the parameter which controls the contribution (or importance) of the weight minimization in the overall objective function ($0 < \mu \leq 1$), and $J(\cdot)$ is the mean square error (MSE) function. For function approximation problem, $J(\cdot)$ can be defined as:

$$J = \frac{1}{L} \sum_{n=0}^{L-1} \sum_{m=1}^{M} \left[y_m(n) - y_m^{NN}(n)\right]^2 \tag{2}$$

where $y_m(n)$ represent the m-th output of the system at time (or training epoch) index n, $y_m^{NN}(n)$ represents the m-th output of the neural network, where $m = 1, 2, \ldots, M$; and L is the total number of training data if bench training is used.

The weight decay term in Eq. (1) can have different quadric forms. For example, an alternative penalty function is used in (9):

$$J_0 = J + \sum_i \epsilon_1 \left(\sum_i \frac{\beta w_i^2}{1 + \beta w_i^2} \right) + \epsilon_2 \left(\sum_i w_i^2 \right) \tag{3}$$

where ϵ_1, ϵ_2 and β are constant coefficients; and w_i represents the connection (weight) vector to hidden unit i.

The drawback with the weight decay method is that it may penalize the weights that have significant influence on the neural network performance. Also, adding the extra penalty term may change the shape of error surface and thus create additional local minima.

Rather than focusing on the magnitude of the weights in the network, the sensitivity based approach attempts to find the contribution of each individual weight in the network and then removes the weights that have the least effect on the objective function. In other words, when the sensitivity of a particular weight is smaller than a pre-set threshold, this weight is insignificant and can be deleted. Furthermore, a neuron can be removed when the sensitivities of all the weights related with this neuron are below the threshold.

The sensitivity s_{ij} of a global error function with respect to a weight can be defined by measuring the difference on the performance of the network with vs. without that weight (12):

$$s_{ij} = J(w_{ij} = 0) - J(w_{ij} = w_{ij}^f) = J(without\ w_{ij}) - J(with\ w_{ij}) \tag{4}$$

where w_{ij}^f is the final value of weight w_{ij} after training. Note that one has to compute Eq. (4) for every weight which is a candidate for elimination throughout all training phases that can be extremely time-consuming for large networks. To simplify, rewrite Eq. (4) as:

$$s_{ij} = -\frac{J(w_{ij} = w_{ij}^f) - J(w_{ij} = 0)}{w_{ij}^f - 0} w_{ij}^f \tag{5}$$

assuming all the other weights (i.e., other than w_{ij}) are fixed (at their final states, upon completion of training). It is suggested that the numerator of Eq. (5) can be calculated as (10):

$$J(w_{ij} = w_{ij}^f) - J(w_{ij} = 0) \cong \int_A^F \frac{\partial J}{\partial w_{ij}} \, dw_{ij} \tag{6}$$

where F is the final state of all the weights on error surface, and A is the state/point when w_{ij} =0 while all the other weights are in their final state.

In Eq. (4), the above integral is calculated along the error surface from a zero-state A to final state F. Note that a typical training process actually starts with some small initial values of NN weights (i.e., state I) which are not equal to (but close to) zero. Therefore, Eq. (6) can be approximated as:

$$J(w_{ij} = w_{ij}^f) - J(w_{ij} = 0) \cong \int_I^F \frac{\partial J}{\partial w_{ij}} \, dw_{ij} \tag{7}$$

This expression can be further approximated by replacing the integral by summation, taken over all the discrete steps (or training epochs) that the network passes while learning. Substituting Eq. (7) into Equations (5) and (6), the estimated sensitivity $\hat{s}_{ij}$ (with respect to the removal of connection w_{ij}) becomes:

$$\hat{s}_{ij} = -\sum_{n=0}^{N-1} \frac{\partial J}{\partial w_{ij}} [\Delta w_{ij}(n)] \frac{w_{ij}^f}{w_{ij}^f - w_{ij}^i} \tag{8}$$

where N is the total number of iterations (training epochs) needed to minimize Eq. (2), and w_{ij}^i is the initial value of weight (before training).

For the back-propagation (steepest descent) algorithm (8), all the weights are updated as:

$$w_{ij}(n+1) = w_{ij}(n) + \Delta w_{ij}(n) \tag{9}$$

and

$$\Delta w_{ij}(n) = -\eta \frac{\partial J}{\partial w_{ij}}(n) \tag{10}$$

where η is the learning rate. Thus, Eq. (8) can be expressed as:

$$\hat{s}_{ij} = \sum_{n=0}^{N-1} [\Delta w_{ij}(n)]^2 \frac{w_{ij}^f}{\eta(w_{ij}^f - w_{ij}^i)} \tag{11}$$

For back-propagation with momentum (8), Eq. (10) becomes:

$$\Delta w_{ij}(n) = -\eta \frac{\partial J}{\partial w_{ij}}(n) + \gamma\, \Delta w_{ij}(n-1) \tag{12}$$

where γ is the momentum constant. Now the sensitivity calculation is slightly different:

$$\hat{s}_{ij} = \sum_{n=0}^{N-1} \left[\frac{\partial J}{\partial w_{ij}}(n)\right] \left[-\eta \frac{\partial J}{\partial w_{ij}}(n) + \gamma\, \Delta w_{ij}(n-1)\right] \frac{w_{ij}^f}{w_{ij}^f - w_{ij}^i} \tag{13}$$

Eq. (11) and (13) indicate that the values of sensitivity are also related with the neural network adaptation law. For different training methods, the way for calculating the gradient $\frac{\partial J}{\partial w_{ij}}$ is the same; but the way for evaluating $\Delta \mathrm{w}_{ij}$ is varied, depending on each individual algorithm. That means, even though we start with the same size of neural network, the same initial value for every weight, the sensitivity matrix may still be distinct if the neural network is trained by different training rules. An appropriate training algorithm can not only improve the training accuracy, but also result in a smaller network as well.

Once the neural network is trained to achieve the input/output mapping with desired accuracy based on certain performance criteria (such as the MSE value), the pruning algorithm can be activated to reduce the size of NN. If the absolute value of the sensitivity of a particular weight is greater than a specific threshold, i.e., the existence of this weight has greater impact on the error function, then it should be kept in the NN weight space; otherwise it can be eliminated:

$$w_{ij} = \begin{cases} 0 & \text{if } |\hat{s}_{ij}| < \tau \\ w_{ij} & \text{if } |\hat{s}_{ij}| \geq \tau \end{cases} \tag{14}$$

Since the partial derivative $\frac{\partial J}{\partial w_{ij}}$ (which is an element in gradient matrix) is always available during training, the only extra computational demand for implementing the pruning procedure is the summation in Eq. (11) (or Eq.(13)); thus is algorithm is easy to implement in practice.

One of the difficulties to implement the above algorithm is to find the appropriate value of threshold in Eq. (14). Ponnapelli et al. (13) suggested that the sensitivities of weights should only be compared with those related with the same node in the same layer. Thus, the concept of local relative sensitivity index (LRSI) is defined as the ratio of the sensitivity of a particular weight and the sum of all the sensitivities of the weights that are connected to the same node from the previous layer:

$$LRSI_{ij} = \frac{|s_{ij}|}{\sum_{m=1}^{M} |s_{mj}|} \tag{15}$$

where M is the total number of connections to node j from the previous layer. For each node, any weight that has a local sensitivity less than a threshold will be pruned:

$$LRSI_{ij} \leq \beta \tag{16}$$

Even though the choice of the threshold (i.e., β) still depends on the rule of thumb, it is now a percentage which is relatively easier to be chosen ($0 \leq \beta \leq 1$), comparing with the selection of τ in Eq. (14).

The pruned NN is then retrained and its performance is evaluated. If the trained NN yields better results than the original network, the pruning procedure can be repeated and a smaller NN can be obtained.

A limitation of the above algorithm is that only weight removal is considered while node pruning is not included. Theoretically, if all the weights that are connected to a single node are pruned, then this node can also be eliminated. In practice, this may take several rounds of pruning and training so it may not be a feasible solution. Also, the network considered in (13) only contains one hidden layer.

Engelbrecht (6) proposed a modified approach to sensitivity analysis from the point of view of statistics. Instead of using the value of the sensitivity directly, Engelbrecht evaluated the average sensitivity of a network parameter (e.g., weight or node) over all the patterns, and then developed a new measure called variance nullity. That is, if the variance of sensitivity of a network parameter over all the patterns (denoted by $\sigma^2_{\theta_k}$ for parameter θ_k) is close to zero and the average sensitivity (also over all the patterns) is small, then we conclude that this parameter has little or no effect on the output of the neural network over all patterns and therefore can be eliminated. The variance of sensitivity is defined as:

$$\sigma^2_{\theta_k} = \frac{1}{P-1}\sum_{p=1}^{P}(s^{(p)}_{\theta_k} - \tilde{s}_{\theta_k})^2 \tag{17}$$

where P is the total number of patterns under consideration, $s^{(p)}_{\theta_k}$ is the sensitivity of θ_k for pattern p, and $\tilde{s}_{\theta_k}$ is the average sensitivity over all the patterns:

$$\tilde{s}_{\theta_k} = \frac{1}{P}\sum_{p=1}^{P}(s^{(p)}_{\theta_k} \tag{18}$$

The parameter variance nullity (PVN) for each parameter is then defined as:

$$\gamma_{\theta_k} = \frac{(P-1)\,\sigma^2_{\theta_k}}{\sigma^2_0} \tag{19}$$

where σ^2_0 is a small constant value related with hypothesis test $H : \sigma^2_{\theta_k} < \sigma^2_0$ (for details, see (6).

Starting from the output layer in a backward order, this algorithm allows the pruning of both nodes and weights in a similar manner layer by layer, with each parameter having a separate formula for the sensitivity calculation. The extension of a sensitivity measurement to nodes (not just weights) allows for the possibility of finding a smaller network, and also decreases the number of times to retrain the network before obtaining its final size. Unlike the algorithms discussed in (10; 13), this algorithm can be applied to NN with multiple hidden layers.

However, as we discussed earlier, relying on one single value of σ_0^2 for the entire network can lead to problems. Fnaiech et. al. (1) suggested that parameters within the same layer should be considered "locally" rather than "globally", and defined a new pruning index called the local parameter variance nullity (LPVN). The PVN for all parameters in the same layer are summed up; then the LPVN for each parameter (which represents the relative importance of PVN of a parameter in the layer) can be obtained and used for pruning:

$$L\gamma_{\theta_k^{[l]}} = \frac{\gamma_{\theta_k^{[l]}}}{\sum\limits_{k=1}^{K} \gamma_{\theta_k^{[l]}}} \tag{20}$$

where $L\gamma_{\theta_k^{[l]}}$ is the LPVN for layer l, and K is the total number of parameters in layer l. Note that in this algorithm, the pruning decision is still based on the hypothesis test H; thus choosing the appropriate threshold for LPVN is crucial to the success of this algorithm.

In practice, the number of weights eliminated also depends on when the pruning process starts. Longer training time before pruning may result in smaller neural network and complete training leads to the minimum neural network size. In addition, if training is continued after pruning, the accuracy of the neural network model can be further improved.

Huynh and Setiono (9) introduced a concept of pruning with cross-validation. The whole dataset is divided into two parts, i.e., the training set T and the cross validation set C. The pruning criterion is still based on the magnitude of each weight (see Eq. (3)); however, a validation step is added to test the pruned network. Pruning is performed on two separate phases, i.e., the weight removal phase and the hidden node removal phase.

At every weight pruning step, the performance of the network with reduced size is compared with the performance of the network before the current pruning phase. Let the performance criterion (objective function) on set T and set C be J_{TR} and J_{CV}, respectively. After pruning, a smaller neural network is obtained and the new performance criterion on set T and set C be J'_{TR} and J'_{CV}, respectively. If

$$(J'_{TR} + J'_{CV}) < (J_{TR} + J_{CV}) \tag{21}$$

i.e., the pruned network outperforms the unpruned one; then the pruned network is accepted and the pruning process can be continued. Otherwise, the network is restored to the size before the current pruning step. Similarly, in the node removal phase, if the removal of hidden node i has the least effect on the pruned network, then it can be pruned. Obviously, the use of an additional cross validation set at each phase of the pruning takes into account that pruning is meant to not only reduce the size of a network, but also improve the network generalization capacity.

As a summary, all the above algorithms have their own advantages and limitations. In the next section, we choose three typical pruning algorithms and compare their performances via computer simulations.

3 Computer Simulation Results

In this section, the performances of some typical pruning algorithms are studied and compared via computer simulations. The three pruning algorithm considered here include the local sensitivity analysis method (KLSA) (13), the local variance sensitivity analysis (LVSA) (1), and the cross validation pruning algorithm (CVP) (9) (for detailed description of each algorithm, see Eq. (15), (16), (20), and (21) in Sect. 2). All three algorithms are tested on the design of the neural network controller for a PS-ZVS-FB (Phase-shifted, zero-voltage switching, full-bridge) DC-DC voltage converter.

As we know, the DC-DC voltage converter plays an important role in many power electronics devices. It is desired that a DC-DC converter keeps its output stable even when the supply voltage and load current fluctuate over a wide range. Under these circumstances, the neural network controller becomes a promising solution based on its adaptive learning ability and nonlinear mapping ability. The output of the neural network, i.e., the control signal for the voltage regulator, is the duty cycle of a square waveform which is used to implement PWM (pulse-width modulation) switching control. The inputs to the neural network include input voltage, load current, and the absolute value of the change of output voltage $|\Delta V_0|$. For a detailed description of the DC voltage regulator and the neural network controller design, see (21),(22), and (23).

The simulation is written in C++ and the general block diagram is shown in Fig. 3. Different initial network configurations are considered to fully test the abilities of each pruning algorithm on a variety of hidden node and hidden layer setups. For this reason, we also extend the KLSA algorithm to the case of NN with more than one hidden layer. Configuration 1 has one hidden layer with fifteen hidden nodes; configuration 2 has two hidden layers with ten nodes in each hidden layer; and configuration 3 has three hidden layers with five nodes in each hidden layer. The data set contains about 1000 input/output data pairs. Similar to the ten-fold cross validation, the data set is divided into ten equal sub-sets; eight of them are used for training, one is used

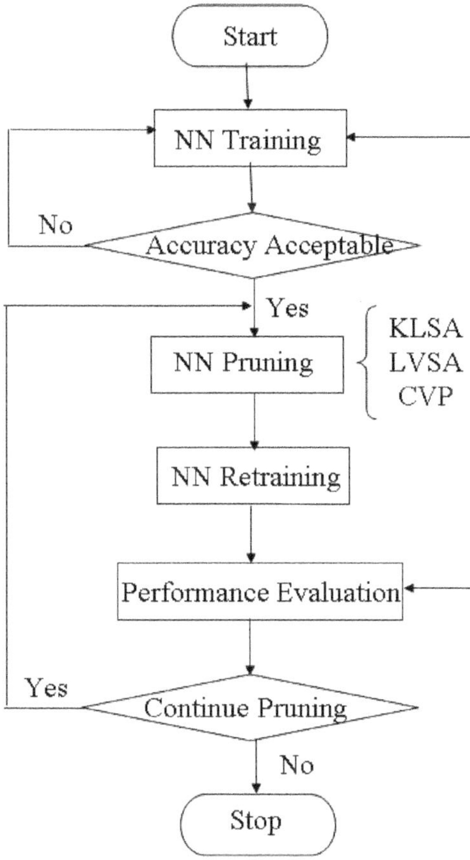

Fig. 3 Computer Simulation

for validation, and the remaining one is for testing. Each of the sub-dataset is used for training, validation, and testing on a rotation basis, resulting in a total of ten different data configurations. During training, all the data are presented in random order. Back-propagation with momentum is an effective training algorithm with fast convergence rate, moderate computational and memory requirement; thus it is chosen to train all the neural networks in this research.

The weights of all the neural networks are initialized to random values before training. The same initial conditions are applied to all the pruning algorithms in each test. To minimize the influences of initial conditions to the test results, ten different sets of initial conditions are chosen for each neural network configuration and each data configuration. Therefore, for each neural network configuration, a total of $10*10 = 100$ simulation runs are performed. This process is repeated for each of the four pruning algorithms.

Table 1 below shows the overall pruning capability of each of the four tested algorithms by displaying the mean and standard deviation (presented as (mean) ± (std)) of the pruning percentage of neural network weights (with respect to the original network). For example, for the first neural network configuration, the KLSA algorithm can prune about 13% of the total neural network weights (an average for 100 runs with different initial conditions and data rotations), with the standard deviation of 12%. Similarly, under the same condition, the CVP algorithm can prune about 34% of the total neural network weights, with a standard deviation of 17%. Obviously, the CVP algorithm outperforms the KLSA algorithm in this case; while the performance of LVSA algorithm is moderate. However, for neural networks with more than one hidden layers, the LVSA algorithm yields the best results. Note that in LVSA algorithm, pruning is based on the relative performance index (LPVN) of different weights that are connected to the same layer; which implies that those weights with smaller LPVN values in each layer can always be removed. However, in CVP algorithm when the network has more than one hidden layer, it may be difficult to relate the removal of the weights that do not connect directly to the output with the neural network performance, because

Table 1 A Comparison of Each Algorithm's Pruning (Percentages)

NN	KLSA	LVSA	CVP
1	13 ± 12	19 ± 13	34 ± 17
2	13 ± 12	23 ± 19	8 ± 21
3	6 ± 5	21 ± 14	5 ± 15

Table 2 The Detailed Results

	KLSA	LVSA	CVP
NN config. 1			
Training	0.0002 ± 0.0001	0.0002 ± 0.0001	0.0002 ± 0.00001
Accuracy	[0.0001, 0.0003]	[0.0001,0.0001]	[0.0001, 0.0003]
Testing	0.0002 ± 0.0001	0.0002 ± 0.0001	0.0002 ± 0.0001
Accuracy	[0.00001, 0.0004]	[0.0001, 0.0011]	[0.0001, 0.0003]
NN config. 2			
Training	0.0002±0.0001	0.0002 ± 0.0001	0.0003 ± 0.0001
Accuracy	[0.0001,0.0004]	[0.0001,0.0003]	[0.0001,0.0011]
Testing	0.0002 ± 0.0001	0.0002 ± 0.0001	0.0003 ± 0.0002
Accuracy	[0.0001,0.0004]	[0.0001,0.0003]	[0.00010,0.0012]
NN config. 3			
Training	0.0003 ± 0.0001	0.0002 ± 0.0001	0.0003 ± 0.0001
Accuracy	[0.0002,0.0005]	[0.0001,0.0004]	[0.0002,0.0007]
Testing	0.0002 ± 0.0001	0.0002 ± 0.0001	0.0003 ± 0.0001
Accuracy	[0.0001,0.0005]	[0.0001,0.0004]	[0.0001,0.0006]

each neuron in the network is a nonlinear MIMO (multi-input, multi-output) system that "interacts" with other neurons in the network.

Table 2 outlines the performance of each algorithm in terms of mean square error on both training and test dataset. The identification error gives a measure of the mean-square-error of the desired output and NN output. It is shown that the overall accuracies of the all three algorithms are similar.

4 Summary

Artificial neural networks have been widely used for many applications in recent years. Before a neural network can be employed, its size must be specified in advance. The purpose of network structure complexity analysis is to find the optimal dimension of a neural network for a specific application.

In this research, three different pruning algorithms are studied, namely, the local sensitivity analysis method (KLSA), the local variance sensitivity analysis (LVSA), and the cross validation pruning algorithm (CVP). The three algorithms are tested and compared on a neural network controller design for a DC-DC voltage converter with different initial network configurations. The simulation results show that the performance of LVSA algorithm is relatively uniform over different network configurations; while the CVP algorithm yields the best pruning result for the network with only one hidden layer. More tests will be conducted to fully evaluate the performances of these three algorithms.

References

1. Fnaiech, N., Abid, S., Fnaiech, F., Cheriet, M.: A modified version of a formal pruning algorithm based on local relative variance analysis. In: First International Symposium on Control, Communications and Signal Processing, pp. 849–852 (2004)
2. Rosin, P., Fierens, F.: Improving Neural Network Generalisation. In: International Geoscience and Remote Sensing Symposium, pp. 1255–1257 (1995)
3. Bevilacqua, V., Mastronardi, G., Menolascina, F., Pannarale, P., Pedone, A.: A novel multi-objective genetic algorithm approach to artificial neural network topology optimisation: the breast cancer classification problem. In: IEEE International Joint Conference on Neural Networks, pp. 1958–1965 (2006)
4. Narendra, K., Parthasarathy, K.: Identification and control of dynamical systems using neural networks. IEEE Transaction on Neural Networks 1(1), 4–27 (1990)
5. Lawrence, S., Giles, C., Tsoi, A.: Lessons in neural network training: overfitting may be harder than expected. In: Proceedings of the Fourteenth National Conference on Artificial Intelligence, pp. 540–545 (1997)
6. Engelbrecht, A.: A new pruning heuristic based on variance analysis of sensitivity information. IEEE Transactions on Neural Networks 12(6), 1389–1399 (2001)

7. Giles, C., Lawrence, S.: Overfitting and Neural Networks: Conjugate Gradient and Backpropagation. In: Proceedings of the IEEE International Conference on Neural Networks, pp. 114–119 (2000)
8. Haykin, S.: Neural networks: a comprehensive foundation. Prentice Hall, New Jersey (1999)
9. Huynh, T., Setiono, R.: Effective neural network pruning using cross-validation. In: IEEE International Joint Conference on Neural Networks, pp. 972–977 (2005)
10. Karnin, E.: A simple procedure for pruning back-propagation trained neural networks. IEEE Transactions on Neural Networks 1(2), 239–242 (1990)
11. Marsland, S., Nehmzow, S., Shapiro, J.: A self-organizing network that grows when required. Neural Networks 15(8-9), 1041–1058 (2002)
12. Mozer, M., Smolensky, P.: Skeletonization: A technique for trimming the fat from a network via relevance assessment. In: Touretzky, D. (ed.) Advances in Neural Information Processing, pp. 107–115 (1989)
13. Ponnapalli, P., Ho, K., Thomson, M.: A formal selection and pruning algorithm for feedforward artificial neural network optimization. IEEE Transactions on Neural Networks 10(4), 964–968 (1999)
14. Yen, G., Lu, H.: Hierarchical genetic algorithm based neural network design. In: IEEE Symposium on Combinations of Evolutionary Computation and Neural Networks, pp. 168–175 (2002)
15. Chang, S.J., Leung, C.S., Wong, K.W., Sum, J.: A local training and pruning approach for neural networks. International Journal of Neural Networks 10(6), 425–438 (2000)
16. Chang, S.J., Sum, J., Wong, K.W., Leung, C.S.: Adaptive training and pruning in feedforward networks. Electronics Letters 37(2), 106–107 (2001)
17. Wan, W., Hirasawa, K., Hu, J., Jin, C.: A new method to prune the neural network. In: Proceedings of the IEEE International Conference on Neural Networks, pp. 449–454 (2000)
18. Neruda, R., Stedry, A., Drkosova, J.: Kolmogorov learning for feedforward networks. In: Proceedings of the IEEE International Conference on Neural Networks, pp. 77–81 (2001)
19. Kamran, F., Harley, R.G., Burton, B., Habetler, T.G., Brooke, M.A.: A fast on-line neural-network training algorithm for a rectifier regulator. IEEE Trans on Power Electronics 13(2), 366–371 (1998)
20. Hecht-Nielsen, R.: Kolmogorov's mapping neural network existence theorem. In: Proceedings of the IEEE International Conference on Neural Networks, pp. 11–14 (1987)
21. Li, W.: A neural network controller for a class of phase-shifted full-bridge DC-DC converter. PhD thesis, California Polytechnic State University, San Luis Obispo (2006)
22. Li, W., Yu, X.: Improving DC power supply efficiency with neural network controller. In: Proceedings of the IEEE International Conference on Control and Automation, pp. 1575–1580 (2007)
23. Li, W., Yu, X.: A self-tuning controller for real-time voltage regulation. In: Proceedings of the IEEE International Joint Conference on Neural Networks, pp. 2010–2014 (2007)

24. Quero, J.M., Carrasco, J.M., Franquelo, L.G.: Implementation of a neural controller for the series resonant converter. IEEE Trans on Industrial Electronics 49(3), 628–639 (2002)
25. Leshno, M., Lin, V., Pinkus, A., Shocken, S.: Multilayer feedforward networks with a non-polynomial activation function can approximate any function. Neural Networks 6, 861–867 (1993)
26. Lin, F., Ye, H.: Switched inductor two-quadrant DC-DC converter with neural network control. In: IEEE International Conference on Power Electronics and Drive Systems, pp. 1114–1119 (1999)
27. El-Sharkh, M.Y., Rahman, A., Alam, M.S.: Neural networks-based control of active and reactive power of a stand-alone PEM fuel cell power plant. Journal of Power Resources 135(1-2), 88–94 (2004)
28. Bebis, G., Georgiopoulo, M., Kasparis, T.: Coupling weight elimination with genetic algorithms to reduce network size and preserve generalization. Neurocomputing 17, 167–194 (1997)
29. Sabo, D.: A Modified Iterative Pruning Algorithm for Neural Network Dimension Analysis. PhD thesis, California Polytechnic State University, San Luis Obispo (2007)
30. Sabo, D., Yu, X.: A New Pruning Algorithm for Neural Network Dimension Analysis. In: Proceedings of the IEEE International Joint Conference on Neural Networks, pp. 3312–3317 (2008)
31. Kopel, A., Yu, X.: Optimize Neural Network Controller Design Using Genetic Algorithm. In: Proceedings of the World Congress on Intelligent Control and Automation, pp. 2012–2016 (2008)
32. Yu, X.: Reducing neural network size for dynamical system identification. In: Proceedings of the IASTED International Conference on Intelligent Systems and Control, pp. 328–333 (2000)
33. Yu, X.: Adaptive Neural Network Structure Based on Sensitivity Analysis. In: Proceedings of the World Forum on Smart Materials and Smart Structures Technology (2007)
34. Brouwer, R.: Automatic growing of a Hopfield style net-work during training for classification. Neural Networks 10(3), 529–537 (1997)
35. Vonk, E., Jain, L., Johnson, R.: Automatic generation of neural network architecture using evolutionary computation. World Scientific Publishing Co., Singapore (1997)
36. Huberman, B., Rumelhart, D.: Generalization by weight elimination with applications to forecasting. In: Lippmann, R., Moody, J. (eds.) Advances in neural information processing III, pp. 875–882. Morgan Kaufmann, San Francisco (1991)
37. Gupta, A., Lam, S.: Weight decay backpropagation for noisy data. Neural Networks 11, 1127–1137 (1998)
38. Reed, R., Marks, R.: Neural Smithing: Supervised Learning in Feedforward Artificial Neural Networks. MIT Press, Cambridge (1999)
39. Goldberg, D.: Genetic Algorithms in Search, Optimization, and Machine Learning. Addison-Wesley, Reading (1989)
40. Bishop, C.: Neural Networks for Pattern Recognition. Oxford University Press, Oxford (1994)

Design and Flight Test of an Intelligent Flight Control System

Tim Smith, Jim Barhorst, and James M. Urnes, Sr.

Abstract. The Intelligent Flight Controls System program is a collaborative effort between Boeing, NASA, small business and academia to implement and flight demonstrate neural-adaptive flight controls technology. IFCS employs neural networks to provide augmentation to the nominal aircraft flight controls in the case of failure conditions to the aircraft. The presence of the neural-adaptive elements in the flight control software presents some interesting challenges in the V&V of the IFCS system. This article will discuss the V&V challenges faced, how they were overcome and what challenges still exist.

1 Introduction

The Intelligent Flight Controls System (IFCS) is a NASA funded, piloted flight test program whose purpose is to demonstrate the ability of neural network technologies to provide compensatory augmentation to a baseline flight control system for adverse flying conditions that may be the result of damage or systems failures to an aircraft in flight. Aircraft flight control software is critical to the safety of flight. Failures of the software to meet the design requirements (e.g. for stability and handling qualities) or to perform as intended could lead to loss of mission, aircraft or human life. The introduction of neural network software poses significant challenges in the design and test of a flight control system. Traditional techniques for validation and verification (V&V) of the flight controller must be amended to accommodate them.

There were two major phases of the IFCS program. In this article we will describe the second generation (Gen II) of the IFCS program in detail. Details of the first generation (Gen I) IFCS program can be found in (1) and (2). Sect. 2 provides an overview of the IFCS program and objectives. Sect. 3 describes the flight test experiment conducted under the IFCS

J. Schumann, Y. Liu (Eds.): Appl. of Neural Networks in High Assur. Sys., SCI 268, pp. 57–76.
springerlink.com

Gen II program. Sect. 4 describes the IFCS Gen II controller architecture. Sect. 5 provides an overview of flight controls requirements validation from the standpoint of aircraft stability and handling qualities and describes how traditional techniques were amended to accommodate the IFCS Gen II neural networks. Sect. 6 describes the software verification issues and methods employed. Sect. 7 describes some flight test results from the IFCS Gen II. Finally, Sect. 8 details some lessons learned from IFCS Gen II.

2 IFCS Program

The purpose of the Intelligent Flight Controls System flight test program is to develop and flight-demonstrate adaptive neural network technologies capable of providing stable flight and adequate flying qualities to an aircraft in the presence of modeling uncertainties, failures and damage to the aircraft. IFCS is a NASA funded, collaborative effort between NASA Dryden Flight Research Center, NASA Ames Research Center, Boeing Phantom Works, the Institute for Scientific Research, West Virginia University and Georgia Tech University. The IFCS program has had two major phases. The second generation IFCS is the focus of this article.

The first generation IFCS was flight tested in 2003. IFCS Gen I employed an indirect adaptive scheme using neural networks to identify and augment stability and control parameters of the aircraft in flight. These parameters provided updates to a linear quadratic regulator flight controller to maintain stability and performance of the aircraft. A pre-trained neural network provided the controller with the healthy aircraft parameters as a function of flight condition and aircraft state. Onboard real-time parameter identification algorithms provided increments to the stability and control parameters which were stored using a dynamic cell structure (DCS), self-organizing feature map routine. The incremented parameters were then passed to the flight controller to optimize performance. The DCS algorithm allowed the flight control system to 'remember' identified parameter increments from previously flown flight conditions so they did not have to be re-learned. The 2003 flight tests successfully demonstrated the parameter identification and retention capabilities of the IFCS Gen I concept. (2) details the IFCS Gen I flight test results.

The second generation IFCS employed a direct adaptive approach to the flight control of the aircraft (3) and (4). That is, adaptive compensation to the controller was based on aircraft tracking performance and sensor measurements without an explicit, on-line estimated aircraft model. The baseline IFCS Gen II controller architecture used a model inverting scheme to provide commands to the aircraft control surfaces that produce the accelerations commanded by the pilot. The neural network architecture is of the Sigma-Pi type. The Sigma-Pi neural networks are described in Sect. 4. (3)

gives a detailed description of the Sigma-Pi implementation used in the IFCS Gen II architecture.

The neural networks interface with the baseline controller by providing augmentation signals to the commanded accelerations when the error regulation portion of the baseline controller fails to provide adequate handling qualities, as determined via measurement of the aircraft state feedback tracking errors signals. One of the features of the IFCS Gen II concept is that the neural network architecture software can be integrated with any conventional flight controller configuration. This offers significant cost savings, especially for retrofit additions to existing controller configurations. The details of the IFCS Gen II controller architecture are provided in Sect. 4.

The IFCS Gen II flight controller was designed to meet the standard requirements for stability and handling qualities of a piloted aircraft. The baseline controller with no neural network augmentation was demonstrated to meet these requirements using standard flight controls development techniques. The addition of the neural networks introduces nonlinearities to the system that complicate the application of traditional stability and performance assessment techniques. Sect. 5 details the requirements validation efforts employed on the IFCS Gen II.

The baseline IFCS Gen II flight control law algorithms are considered flight critical software. They were tested to NASA Class A standards (5). The Sigma-Pi neural network algorithms are considered mission critical software and were therefore tested to NASA Class B standards. To ensure system integrity while concurrently running software of different criticality classes, the baseline controls and neural networks were run on separate processors interfaced with a MIL-STD-1553 data bus. Safety monitors, hosted on the Class A processor, provided absolute and rate limiting of the neural network augmentation signals. In this article, the terms 'Level' and 'Class' are used interchangeably in reference to software criticality classification. Sect. 6 describes the software and systems verification efforts employed on IFCS Gen II.

Flight tests of the IFCS Gen II began in 2006 and continued into 2008. IFCS flight test results are summarized in Sect. 7.

3 IFCS Experiment

The IFCS testbed aircraft, shown in Fig. 1, is the NF-15B tail number 837. The aircraft is a B model F-15 that was extensively modified by the McDonnell Douglas Corporation and the U.S. Air Force to serve as a Short Take Off and Landing Maneuver Technology Demonstrator (STOL/MTD) (6). Modifications to the aircraft include the addition of canard flight control surfaces, a fully digital fly-by-wire flight control system and axisymmetric thrust vectoring nozzles. The thrust vectoring nozzles were not used as control effectors for the IFCS program.

Fig. 1 NASA NF-15B Tail No. 837

To demonstrate the ability of the neural networks to provide stabilizing control augmentation and adequate handling qualities to a damaged aircraft, simulated failure modes were incorporated into the IFCS Gen II flight controls logic. The failure modes are pilot selectable through a system called Dial-A-Gain (DAG) and Choose-A-Test (CAT) whereby the pilot can choose the failure type and magnitude through button presses on the cockpit digital display indicator. There are two basic failure modes incorporated into the IFCS control logic. The two simulated failure types consist of a change to the baseline collective canard schedule and a lock-in-place of the left stabilator. These are referred to as the 'A matrix' and 'B matrix' failures respectively.

For a given mach and altitude, the collective canard deflection on the IFCS aircraft is commanded to a value that is scheduled with the angle of attack. The canard schedule effectively controls the air flow over the wing and provides a partial lift augmentation to the aircraft. The IFCS Gen II experiment allowed for a selectable gain to be placed in the collective canard command path that could be altered between values in the range [-1.75, 1.0]. A gain of 1 is the nominal configuration. When a simulated canard failure is induced, the gain is altered to an off-nominal value which changes the lifting characteristics of the aircraft. This type of simulated failure is termed an 'A matrix' failure because in the linear representation of the aircraft dynamics, variation of the collective canard command gain alters the aerodynamic derivatives stored in the state (colloquially A) matrix. This type of simulated failure corresponds to a symmetric lift loss and degradation of pitch damping – as if the aircraft had lost an equal percentage of both wings. While this type of failure would not typically be expected for aircraft, the symmetry of the simulated failure confined the change in aircraft dynamics to the longitudinal axis and allowed for a clear evaluation of the adaptive capabilities of the neural networks in a single axis.

The second type of simulated failure demonstrated by the IFCS Gen II experiment is lock-in-place of the left stabilator. When this type of failure is selected, the left stabilator is commanded to a value between plus and minus four degrees from the current trim value of the stabilator. For example, when a four degree stabilator lock failure is selected, the stabilator is driven to a position four degrees from the current trim value and locked in place. When zero degrees is selected, the stabilator is command to remain at the current trim value for the duration of the experiment. This failure is called a 'B matrix' type failure because in the linear representation of the aircraft dynamics, locking the stabilator affects the control derivatives in the control (colloquially B) matrix of the representation. Since only a single control effector is locked, the resulting aircraft dynamics become highly coupled between the longitudinal and lateral-directional axes. This type of simulated failure provides a more realistic failure scenario for aircraft, as compared to the simulated canard failure, and tests the ability of the neural networks to provide adequate compensation to all three aircraft axes simultaneously.

The IFCS Gen II control laws have three operational modes. These are:

- baseline research flight control mode
- research flight control mode with simulated failure but neural networks inactive
- research flight control mode with simulated failure and neural networks active

In baseline research flight control mode the research flight control laws are active but no simulated failures to the canards or stabilator are introduced and the neural networks are not providing augmentation to the controls. This is the nominal behavior of the IFCS Gen II research control laws. To introduce a simulated failure experiment the pilot can select from several values of the DAG; each of which correspond to either no failure, a canard schedule gain change of various value or a left stabilator lock-in-place to several values, as described above. The CAT function was used to select whether the neural networks would be on or off for a given experiment. The ability to select whether simulated failures and neural networks are active or inactive is an important aspect of the IFCS Gen II experiment. These options allow the pilot to compare the handling qualities of the aircraft without failure or neural networks, with failure but no neural networks active, and with failure and neural network augmentation active. The pilot can then provide an assessment of how handling qualities degrade with surface failure insertion and whether or not and by how much they improve with the neural networks providing compensation.

4 Controller Architecture

The IFCS Gen II controller architecture consists of the baseline research flight controller and Sigma-Pi neural networks. A separate neural network was implemented in each of the three axes (Longitudinal, Lateral and Directional) in the Gen II controller. As mentioned above, the baseline research flight controller can operate with the neural networks active or inactive. First, we detail the baseline IFCS Gen II controller architecture and then we describe the neural network architecture and its interaction with the baseline. Fig. 2 provides an overview of the IFCS Gen II system architecture.

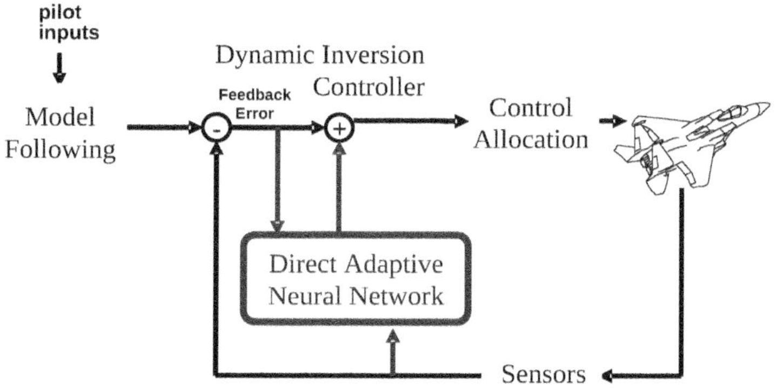

Fig. 2 IFCS Gen II System Architecture

The IFCS Gen II flight controller uses a dynamic model inverting control scheme with Proportional, Integral, Derivative (PID) regulation of the errors between the pilot commands and the sensed aircraft response. The pilot's stick and pedal commands are converted into desired aircraft angular rates and accelerations for the three aircraft axes (longitudinal, lateral and directional) in the Model Following block of Fig. 2. The PID error regulator computes tracking error compensation signals that are summed with the desired angular accelerations. These signals are the total desired aircraft angular accelerations. We will denote them in vector form as $\dot{x}$, where the dot indicates differentiation with respect to time. A low order approximation to the dominant airframe dynamics can be written in the form of a first-order differential equation:

$$\dot{x} = Ax + Bu \quad \Rightarrow \quad u = B^{-1}(\dot{x} - Ax) \tag{1}$$

Where **A** and **B** are, respectively, the state and control matrices mentioned above, **x** is the vector of aircraft angular velocities and **u** is the vector of control inputs to the aircraft. We can then solve for the desired control inputs that provide the desired angular accelerations in the manner shown to the right of the arrow in the above equation. This process is the basis of the IFCS Gen II model inverting control contained in the Dynamic Inversion Control block in Fig. 2. The airframe dynamics parameters contained in the **A** and **B** matrices are dependent on the flight condition and are stored in the controller in the form of a table lookup routine. The Control Allocation block in Fig. 2

contains logic that parcels the calculated commands to the individual flight control effectors.

The neural network architecture chosen for the IFCS Gen II system is of the Sigma-Pi type. This architecture was chosen based on the results of a trade study conducted to compare the accuracy and adaptation speed of multiple neural network architectures. The Sigma-Pi neural networks provide adaptation to the IFCS Gen II controller by adding augmenting signals to the total desired aircraft accelerations described above. Fig. 3 shows the neural network architecture.

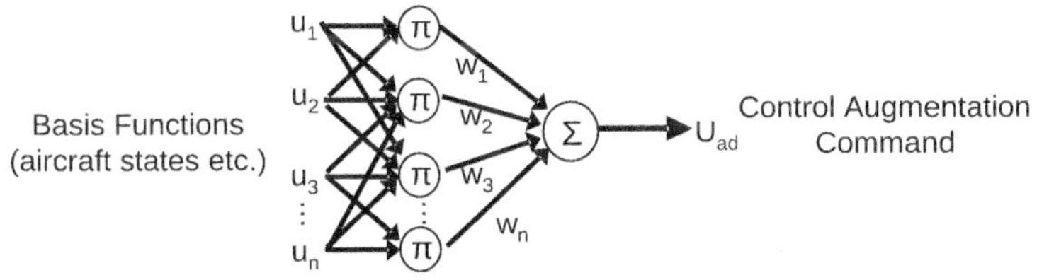

Fig. 3 Sigma-Pi Neural Network Architecture

The Sigma-Pi network is so-called because it calculates the weighted sum of basis functions. For IFCS Gen II these basis functions, u_i, are sensor measurements and functions of the tracking errors in the three aircraft axes. The signal U_{ad} is the augmenting signal that is provided to the baseline controller to compensate for tracking errors induced by the simulated surface failures. The neural network weights, w_i, are adaptive. The adaptive laws that govern the evolution of the neural network weights are derived from Lyapunov nonlinear systems stability theory and ensure bounded command tracking while guaranteeing that the system remains stable (7). The details of the weight adaptation laws are discussed in Sect. 5.4 below. The weight adaptation laws are also equipped with dead zones that stop adaptation when the command tracking errors are within certain, acceptable tolerance levels. The dead zones are necessary to avoid a phenomenon known as parameter drift that can lead to sudden catastrophic system instability for adaptive systems (8). The details of linear and nonlinear stability analysis are covered in the next section.

5 Requirements Validation

One of the most important aspects of a control systems design is the process of validation that the design meets the overall system requirements. Three of the major requirements concern quantification of system stability, aeroservoelastic margin and aircraft performance (or handling qualities). Each of these requirements is critical for control systems design and the approach to meeting each of these requirements for the IFCS Gen II control system had to be amended to accommodate the neural network algorithms. In this section we will describe each of these requirements, how they are met for traditional control systems and how the challenges to validation posed by the presence of the neural networks were overcome for the IFCS Gen II controller.

5.1 System Stability

System stability is the tendency of a system to return to an equilibrium state when perturbed from that equilibrium. For an aircraft, this means that if the aircraft is in steady flight (i.e., with zero rotational rates) and it is perturbed via a temporary pilot command or disturbance, then it will return to steady flight after some settling period. Stability robustness margins provide quantitative measures of the strength of this tendency to return to equilibrium. These measures are gain margin and phase margin. The gain and phase margins, respectively, quantify the amount of gain or phase uncertainty can be withstood by the system without the system becoming unstable. Typical requirements for these margins are 6 dB of gain margin and 45 degrees of phase margin. This means that for any single control loop an artificial gain equivalent to 6 dB or a time delay equivalent to 45 degrees of phase lag can be inserted while the system remains stable. These margin values are measured using a linearized model of the aircraft and can be read from the Bode frequency response of the open loop transfer function with each of the control loops broken one at a time. Fig. 4 depicts the calculation of the gain and phase margins.

The gain margin is measured as the distance, in dB, from the 0 dB line to the magnitude frequency response curve at the frequency at which the phase response crosses 180 degrees for the given control loop. The phase margin is measured as the distance from the 180 degrees line to the phase response curve at the frequency at which the magnitude response crosses the 0 dB line. These margins have to be evaluated and the requirements met for each control loop in the system and the analysis must be performed over the entire flight envelope. (i.e. Mach number and altitude combinations) For the baseline Gen II IFCS controller, without neural networks in the loop, the stability margin requirements were met throughout the IFCS Gen II flight test envelope by tuning the gains in the PID error regulator. The tuning process is described in Sect. 5.3 below.

5.2 Aeroservoelastic Margin

A typical aircraft structure has natural vibration mode frequencies in the bandwidth of the flight controls. Feedback signals with frequency content in that bandwidth must be attenuated to prevent the controller from commanding the effectors to operate at these frequencies, which could excite the

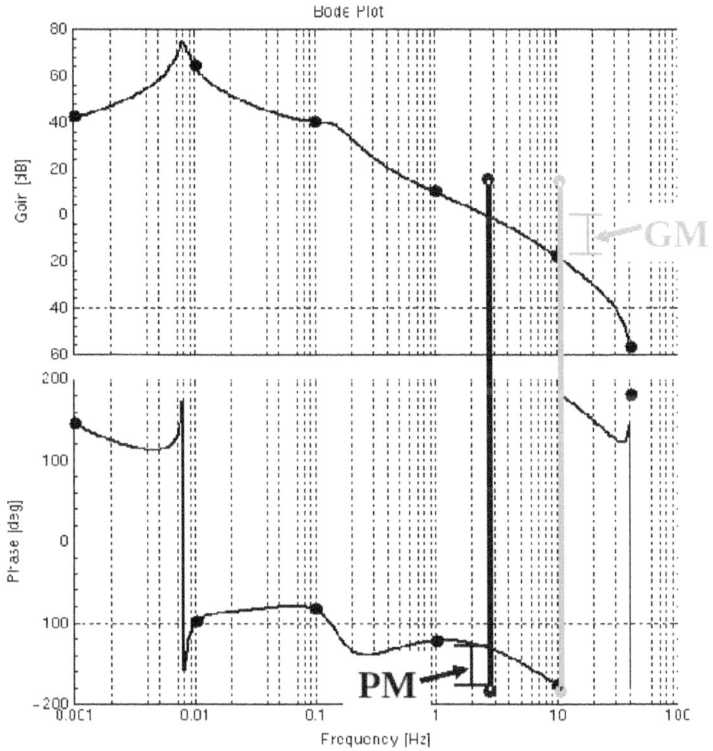

Fig. 4 Gain and Phase Margin Calculation

aircraft's natural vibration modes. If the controller were allowed to output commands in the frequency range of the aircraft's structural mode frequencies, it could cause the natural flexure of the aircraft to be interpreted by the sensors as uncommanded disturbances that the controller would try to regulate by commanding more deflection in the frequency range of the flexible structural modes. This creates a positive feedback loop that could literally shake the aircraft apart. To avoid this phenomenon, bending mode compensation filters are applied to the feedback signals to attenuate them in the frequency range of the aircraft structural modes. Typical requirements for Aero Servo Elastic (ASE) margin are that the open loop frequency response must demonstrate 8 dB of attenuation for frequencies at and above the first structural mode frequency of the aircraft. This analysis is again performed in the frequency domain on a linearized model of the aircraft. The ASE margin is measured as the distance from the 0 dB line to the frequency response curve at the lowest bending mode frequency for the aircraft. Fig. 5 demonstrates this measurement.

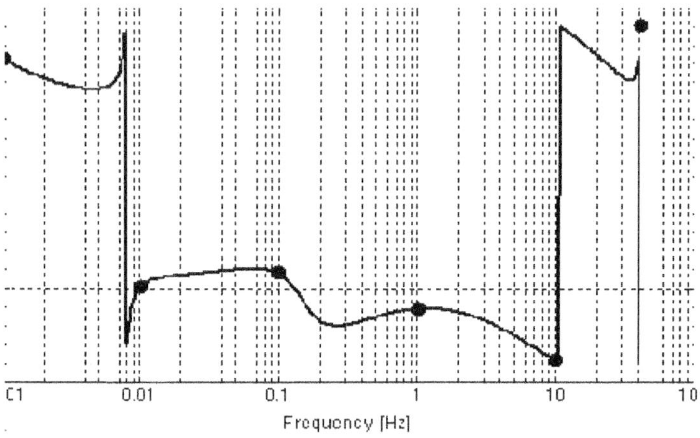

Fig. 5 Aeroservoelastic Margin Measurement

For the IFCS Gen II controller, again without neural networks, the ASE margin requirements were assessed and met by altering the bending mode compensation filters and the PID error regulation gains.

5.3 Handling Qualities

Handling qualities for an aircraft are assessed using both linear and nonlinear analysis techniques. Once a candidate controller is designed, the aircraft is simulated using a nonlinear, six degree of freedom simulation routine. Standard test inputs are applied (such as a longitudinal stick doublet) and the responses saved. From the saved response, equivalent linear model parameters can be measured and compared with standard requirements such as those called for in MIL-STD-1797A, *Flying Qualities of Piloted Vehicles* (9). These handling qualities analyses were employed in the design of the IFCS Gen II flight controller.

These three requirements (stability, ASE and handling qualities) are somewhat contradictory and require tradeoffs in the design of the controller to simultaneously meet them. For instance, it is possible to achieve very fast response to pilot inputs by implementing very large gains in the PID error regulator, but this could lead to insufficient damping of the response and would violate the stability requirements. Also, ASE filtering typically introduces phase lag in the control loops that can adversely affect the phase margins. The design process is an iterative one whereby the controller gains are tuned to meet, say, the stability requirements. Then the ASE margin and performance are assessed. The gains are adjusted to meet these requirements and the process continues in a loop until all requirements are simultaneously satisfied. This process is demonstrated pictorially in Fig. 6 below.

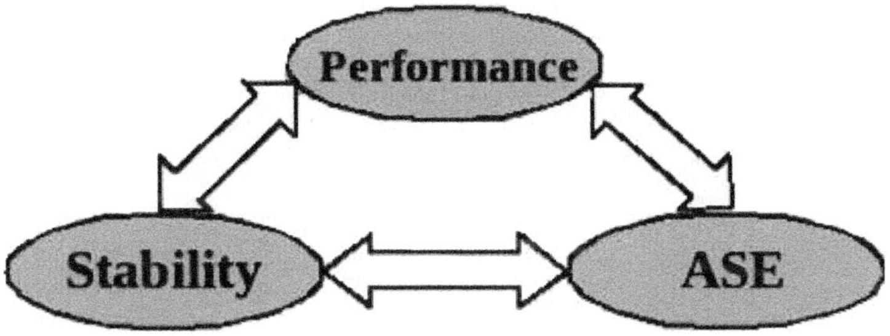

Fig. 6 Iterative Controller Design Process

5.4 Nonlinear Systems Requirements Validation

The validation analyses above all rely on linearized aircraft models. For the baseline IFCS Gen II flight controller design these methods were employed directly to assess the requirements validity. In this way, the IFCS Gen II flight controller requirements, with no neural networks active, (both with and without simulated failures) were validated. This wasn't possible with the neural networks in the loop because the time varying nature of the adaptive weights in the Sigma-Pi architecture does not lend itself to linearization. In order to show the stability of the system with the neural networks active, nonlinear systems stability theory was used. In particular, Lyapunov's second method provides the needed theoretical constructs.

Lyapunov's second method is essentially a mathematical formalization of the intuitive notion that a system that constantly dissipates its stored energy must eventually settle to an equilibrium state. If there exists V, a positive definite scalar function of the states of a system whose time derivative is negative definite, then V is said to be a Lyapunov function for the system. If such a function can be found, then the system is guaranteed to be stable in the sense of Lyapunov. The Lyapunov function used to show the stability of the IFCS Gen II controller *with* neural networks in the loop is shown in Equation 2 below.

$$V = e^T Pe + \mathrm{trace}(\Delta W^T \Gamma^{-1} \Delta W|\Lambda|) \qquad (2)$$

Where

- e is the tracking error
- P is the unique PD matrix defined by $PA + A^T P = -Q$ $(Q = Q^T > 0)$
- W is the error in the NN weights
- Γ is a symmetric PD matrix
- Λ is a diagonal matrix of aerodynamic uncertainties

The neural network weight adaptation law (with dead zone) used for the IFCS Gen II neural networks is shown in Equation 3 below.

$$\dot{W} = \begin{cases} \Gamma\Phi(x)e^T PB \ \mathrm{sgn}(\Lambda) & ||e|| > \epsilon \\ 0 & ||e|| \leq \epsilon \end{cases} \tag{3}$$

Where

- W is the NN weight vector
- $\Phi(x)$ is the vector of NN basis functions
- B is the (known, up to multiplactive uncertainty) control matrix
- ϵ is the radius of the dead zone ball about the origin

(8) gives the details of the derivation of the weight adaptation law given in 3. It is shown that for this weight adaptation law, the above function does indeed have negative semi-definite time derivative. Furthermore, application of Barbalat's Lemma (10) gives that the tracking errors remain bounded and the Sigma-Pi weights converge to their ideal values as time approaches infinity. It is noted that knowledge of neither the true, uncertain (due to simulated failures) state (A) matrix for the damaged aircraft or the neural network approximation error (ΔW) is required. The solution to the Lyapunov equation (P) is generated using the (known) state matrix for the undamaged aircraft. As with linear stability analysis, an accurate linear model of the undamaged aircraft is required for the nonlinear analysis. Both linear and nonlinear stability analysis techniques rely on the assumption that all significant aircraft dynamics are modeled accurately in the underlying linear models.

Nonlinear stability theory via Lyapunov functions gives us that the system, with neural network augmentation, is guaranteed to be stable, but that is not the whole picture. Our requirements are that stability must be shown to be robust. Unfortunately, Lyapunov theory does not address stability robustness. Nor does it provide insight into the ASE margin analysis. To address this shortcoming in the theory and still show that the IFCS Gen II system was safe to fly, advantage was taken of the facts that the simulated failure modes for the IFCS experiment were known a priori and that, for a given set of neural networks weights, the Sigma-Pi neural networks are indeed linearizable. For each of the simulated failure modes and experiment flight conditions, nonlinear simulations were made of the aircraft experiencing the simulated failure followed by pilot input designed to train the neural networks. These neural network training inputs were applied until the Sigma-Pi neural network weights converged. The evolution of the Sigma-Pi weights was recorded during these simulations. The recorded weights were then input to a linear model of the aircraft; which, as mentioned above, is now possible to achieve due to the weights being 'frozen' at a particular time instant. The linear models with the frozen neural network weights were created for each of the experiments at intervals of three seconds. These models were used to demonstrate that the stability robustness and ASE margin requirements were indeed met as the neural network weights evolved toward their converged values. Since Lyapunov stability theory guarantees the convergence of the neural network weights, the handling qualities requirements validation was facilitated by performing this analysis on the nonlinear simulation with these converged weights. The handling qualities metrics for the time intervals during which the neural network weights were converging were not assessed since they would be constantly changing as the neural network adapted. It was deemed sufficient to demonstrate through simulation that the aircraft did not display tendencies toward instability during this period of transient weight adaptation. Fig. 7 shows the evolution with time of the gain and phase margins for the system with the left stabilator locked at a position four positive degrees from trim as the neural network weights converged.

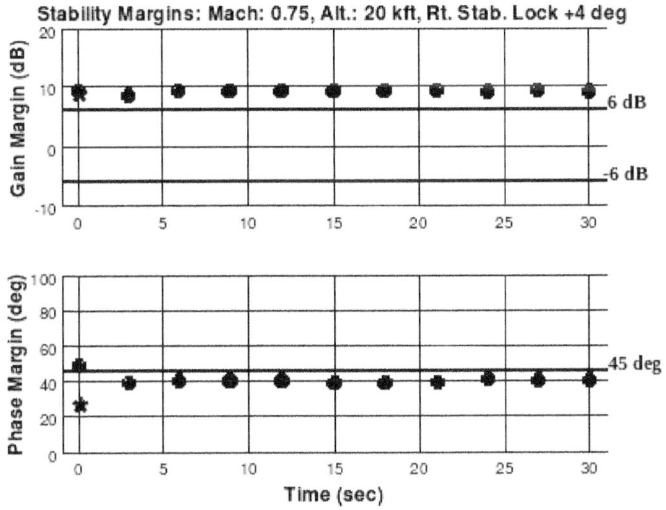

Fig. 7 Stability Robustness Evolving With Time

The dots represent the gain and phase margin measured at the given time instant. The stars at time 0^+ represent the margins immediately after insertion of the simulated failure and before the neural networks have had time to react. The star in the gain margin plot of Fig. 7 almost coincides with the dot, indicating that the stabilator lock failure had little effect on the gain margin of the system before the neural network augmentation was applied. The gain margins for this experiment never dropped below the required 6 dB value. The phase margins did exceed the 45 degree limit but they recovered to roughly 40 degrees of phase margin quite rapidly. NASA's Flight Readiness Review Board accepted that this slight infringement of the phase margin requirement was acceptable for flight test.

The process of simulating the adaptive neural network weights and analyzing the stability and ASE margins using traditional linear analysis means is obviously not ideal because the validation methods employed relied on known failures inserted to the aircraft followed by known pilot inputs for neural network training. In a real world situation the failures affecting an aircraft would most surely not be known a priori and the pilot inputs following the failure would depend on many factors, including the personality of the pilot, weather conditions, air traffic, etc. Although the stability and ASE margins for the IFCS Gen II experiment were demonstrated as adequate to safely fly the test flights, these methods will need refinement before transitioning to a production aircraft. Recent research in the area of gain and time delay margins for some nonlinear, adaptive control architectures is yielding very promising results. For example, in (11) Cao et al. describe an extension to the controller architecture employed by the IFCS Gen II that has guaranteed gain and time delay margins. Time delay margins are the nonlinear analog to phase margins.

6 Flight Controls Software and System Verification

In addition to the validation discussed in Sect. 5, verifying that the software has been properly implemented (in effect, meeting all of the system's software and design requirements) is equally important. A run-time error or unhandled software exception can lead to unexpected system behavior including total computer shut down; which could obviously lead to tragic consequences in a flight critical system. However, verification is more than simply testing the code for "bugs".

Proper verification of software consists of a number of elements, including requirements traceability, some type of code review, unit testing, integration testing, and system testing. System requirements are often hierarchical in nature and are crucial for testing and verification through their traceability throughout development. A code review by someone other than the developer can significantly reduce the number of errors. Unit testing, performed on the individual code modules, serves two purposes: low level functional verification and structural (full path/statement coverage) verification. Properly done, unit testing not only ensures that the module's outputs are correct as a function of normal inputs (i.e., meeting the lowest level requirements), and full coverage (structural) testing ensures that there is no possibility that code in the unit will cause a failure, even when subjected to inputs not explicitly seen during requirements testing. Integration testing takes these successfully tested units, and tests a group of related units together (against "mid-level" requirements), now focusing on their interfaces. System testing (of the highest level requirements) is performed on the fully integrated system, performing an end-to-end checkout, and is as representative of the actual flight system as possible.

Verification and validation are often used as a means of gaining airworthiness certification. Certification is the process for obtaining regulatory approval (from an agency) for an air vehicle by establishing that it meets all applicable governmental regulations. Certification is a means to establish a level of confidence in a system. Certified systems are expected to perform more reliably than uncertified ones. The United States Federal Aviation Administration (FAA) controls commercial and private air vehicle certification. To maintain the public's confidence in safety, the FAA maintains a comprehensive certification process. The following list of advisory circulars and notices governs the FAA's certification of software for use in air vehicles.

- Advisory Circular 20-115B, RTCA, Inc., Document RTCA/DO-178B
- Advisory Circular 21-33, Quality Assurance of Software Used in Aircraft or Related Products
- Advisory Circular 21-35, Computer Generated/Stored Records
- Advisory Circular 21-36, Quality Assurance Controls for Product Acceptance Software
- Notice 8110.85, Guidelines for the Oversight of Software Change Impact Analyses used to Classify Software Changes as Major or Minor
- Notice 8110.86, Guidelines for Software Conformity Inspection and Software Conformity Review
- Notice 8110.87, Guidelines for Determining the Level of Federal Aviation Administration (FAA) Involvement in Software Projects
- Notice 8110.89, Guidelines for the Approval of Software Changes in Legacy Systems Using RTCA DO-178B (formerly N8110.78)
- Notice 8110.90, Guidelines for the Software Review Process (formerly N8110.81)
- Notice 8110.91, Guidelines for the Qualification of Software Tools Using RTCA/DO-178B (formerly N8110.83)
- Notice 8110.92, Guidelines for Applying the RTCA/DO-178B Level D Criteria to Previously Developed Software (PDS) (formerly N8110.82)
- Notice 8110.93, Guidelines for the Approval of Field-Loadable Software by Finding Identicality Through the Parts Manufacturer Approval Process (formerly N8110.79)
- Notice 8110.94, Guidelines for the Approval of Airborne Systems and Equipment Containing User-Modifiable Software (formerly N8110.84)
- Notice 8110.95, Guidelines for the Approval of Field Loadable Software (formerly N8110.77)
- Notice 8110.97, Guidelines for Approving Reused Software Life Cycle Data

The key document that gets all the attention is DO-178B, *Software Considerations in Airborne Systems and Equipment Certification* (12) (the first bullet above). It contains the bulk of the guidelines that the FAA currently follows for certification. DO-178B presents objectives for software life cycle processes, descriptions of activities and design consideration for achieving those objectives, and descriptions of the evidence that indicates that the objectives have been satisfied. It characterizes failure conditions ranging from catastrophic to hazardous/severe-major, to major, minor and no effect. It then defines software levels based on the contribution of software to potential failure conditions from Level A – software whose failure would cause a catastrophic failure of the aircraft, through Level E – software whose failure would have no effect on the aircraft's operational capabilities. The type and rigor of testing for each level is thoroughly defined. Fundamentally, there are two complementary types of testing – requirements testing and structural (or coverage) testing. Requirements testing ensures that all of the requirements have been properly implemented. In flight and safety critical applications, all the code needs to be tested – even if it is not normally executed in fulfilling the requirements. This is called structural or coverage testing and there are several ways it can be done depending on the criticality of the application. There are three basic types of coverage testing (from least rigorous to most rigorous):

- Path – Every statement executed at least once
- Branch – Every branch taken at least once
- MCDC – Every possible combination of logicals presented at each branch

NASA has developed its own approach to software assurance. This is contained in (5), NASA document DCP-S-007 "Software Assurance" from the Dryden Flight Research Center where 3 levels of software classification are defined:

Level A: Software failure could cause loss of life, life-threatening injury, compromise public safety, or result in loss of or substantial damage to the vehicle/system/facility.
Level B: Software failure could cause loss of flight research mission/test.
Level C: Software failure could cause inaccurate results or inefficient use of resources.

Project software assurance activities are defined in the document including software V&V, configuration management, software quality assurance, and software safety.

As mentioned above, the software in the IFCS Gen II baseline research flight control software was designated as NASA Level A, and it was tested accordingly. Coverage testing to Level A standards was performed and each software module was independently peer reviewed. Peer reviews, where someone other than the software module developer examined the code for accuracy and system's requirements compliance, are an important aspect of software verification and found to be worthwhile on IFCS. The baseline flight control software was hosted on a separate processor from the Level B designated Sigma-Pi neural network software.

Verification of the Sigma-Pi neural network software posed a new challenge for verification testing. In the past, software developers have usually written and verified fully deterministic software that is relatively straightforward to understand and predict its execution sequence. This is an opposite of nondeterministic code that does truly random, unpredictable things and seems to be unverifiable. For example, nondeterministic code might rely on a random number generator. But there also is another type of which the IFCS Sigma-Pi neural network is an example. This is code that only appears to be nondeterministic due to its complexity. Its operation (in this case the neural network learning) may be a function of the time histories of a number of system inputs; fully deterministic yet beyond the capacity for a person to fully understand and analyze. Thus it appears to be nondeterministic and unverifiable. But with the right type of tools and analysis, safe operation of the code can be verified.

On IFCS, the testing of the Sigma-Pi neural network was an instance of this. Sigma-Pi neural networks are learning neural networks. Learning neural network software architectures pose a new and different problem for certification processes and for the military and civil certification requirements. Both agencies are very interested in software certification process improvements to deal with adaptive control and learning systems. From a verification standpoint, a learning neural network is generally perceived by the software certification community to be nondeterministic. It was decided to employ a safety monitor at run-time to limit the output authority of the neural network and thus treat the Sigma-Pi software as Level B. The safety monitor software was resident on the Level A processor and ensured the integrity of the data received from the Level B processor hosting the neural network software before allowing it to augment the IFCS Gen II flight controls. The safety monitor was implemented as a floating limiter that limited both the rate and magnitude of the Sigma-Pi outputs. Rate limiting of the Sigma-Pi outputs provided assurance that unexpected, highly dynamic data would not compromise the safety of the aircraft. Absolute limiting ensured that the authority of the augmentation signals was within preset bounds determined by extensive analysis and simulation. Violation of the safety monitor limits initiated a reversion from the IFCS Gen II research mode to conventional F-15 flight control laws. Path coverage testing was performed on the neural network software. This approach to verification allowed the Class B software to interact safely with the Class A baseline controller and was acceptable to NASA for the IFCS Gen II flight tests.

Subsequent analysis showed that, although complex, the software implementation of the Sigma-Pi neural network was fully deterministic. The same "training" set on the neural network always produced the same result. This is not surprising since the Sigma-Pi software implementation does not contain any software constructs not found in flight controller implementation. There are no explicitly time dependent or random elements in the software. While the same input set applied to the neural network will always produce the same results, there is some uncertainty in characterizing the effect on the neural network output resulting from an incremental change in the input set. Traditional software verification techniques coupled with safety monitoring and rigorous piloted and unpiloted simulation demonstrated that the Sigma-Pi neural network implementation was certifiable to NASA Level B standards and the IFCS Gen II system was deemed acceptable for flight test by the NASA Flight Readiness Review Board.

7 Flight Test

Flight tests for the IFCS Gen II system began in 2006 and continued into 2008 (13) and (14). The flight tests were conducted at NASA Dryden Flight Research Center and were flown by two NASA test pilots. During a typical flight test experiment, the pilot evaluated the performance of the nominal aircraft operating under the IFCS Gen II baseline control laws by performing a set of test maneuvers. Once the pilot had assessed the baseline control response, the simulated canard or stabilator lock failure was inserted and the pilot assessed the degradation in response due to the failure by performing the test maneuvers a second time. Then the pilot activated the neural network algorithms, performed some training inputs to get the neural network weights to converge, and then performed the test maneuvers a third time and assessed whether or not and by how much the handling qualities of the aircraft had improved due to the neural network control augmentation.

The canard schedule change and stabilator lock-in-place simulated failures were evaluated by both pilots. The handling qualities metrics evaluated were Gross Acquisition (GA), Fine Tracking (FT) and pilot induced oscillation (PIO). Gross acquisition refers to the pilot workload required to place the targeting reticule on a target, which may be another aircraft performing maneuvers. Fine tracking refers to the pilot workload required to keep the targeting reticule on the target. PIO refers to the tendency of the control laws to produce commands that cause the aircraft motion to be sufficiently out of phase with the pilot inputs that oscillatory motion of the aircraft is induced. Handling qualities were assessed using the Cooper-Harper Rating (CHR) system. The CHR scale takes values from one to ten with one being 'highly desirable' and ten indicating 'major deficiencies'. The CHR scale is described in detail in (9).

Table 1 summarizes some of the results presented in (13) and relates the handling quality assessments for a canard schedule gain of -0.5 times the nominal value for a formation flight task. Results for both pilots A and B are presented. Pilot B gave separate PIO ratings for the GA and FT tasks for the neural networks on configuration.

Pilot A felt that the introduction of the simulated failure to the control system degraded aircraft's handling qualities for GA, FT and PIO. He further

Table 1 CHR for Canard Schedule Change Simulated Failure

Canard Multiplier (Pilot A / Pilot B)	Gross Acquisition (GA)	Fine Tracking (FT)	PIO scale rating
Baseline; formation flight	2/2.5	2/1.5	1/1
-0.5 Canard Gain; NN off; formation flight 2.5	2.5/3	3/4	4/3
-0.5 Canard Gain; NN on; formation flight	2/4	2/2	1/GA:3 1/FT:1

felt that the neural network augmentation improved the response so that it was close to the baseline aircraft. Pilot B also indicated a degradation of CHR when the simulated failure was inserted. Pilot B felt that the GA performance was slightly degraded with the neural networks active but that the FT performance was improved. Pilot B felt that the PIO tendency for GA was not affected by the neural networks but that PIO for FT was improved significantly.

Both test pilots also flew experiments with the stabilator lock-in-place simulated failure. Both pilots noted significant degradation of the aircraft handling qualities with the insertion of the failure. Over the range of different simulated failure magnitudes, the pilots noted slight improvement to significant degradation of the handling qualities when the neural networks were activated. The stabilator lock-in-place simulated failure is asymmetric in nature and induces significant coupling between the longitudinal and lateral-directional axes. The neural networks were not able to completely remove this coupling due in large part to the dead zones on the neural network weight adaptation. However, as noted above, the dead zones are necessary to prevent parameter drift and to ensure that the system does not become over sensitive. (14) provides a detailed analysis of the IFCS Gen II flight test results.

8 Conclusions

The Intelligent Flight Control System project demonstrated in flight test the ability of adaptive neural network technology to compensate for adverse flight conditions such as those that could occur due to systems failures or structural damage. The controller architecture discussed herein is the second generation of the IFCS program which used a model inverting baseline controller with PID error regulation and Sigma-Pi neural network elements to augment the flight controls in the event of a simulated failure to the aircraft. The addition of neural networks to the flight controls posed new challenges in the validation and verification of the IFCS system. The application of nonlinear systems stability theory coupled with traditional flight controls validation techniques allowed for the validation that the flight control laws met the system requirements. While the validation methods employed are not likely to suffice for a production program employing neural network technologies, recent research has made important steps toward comprehensive nonlinear stability robustness measurement techniques. The baseline controller software was verified as NASA Class A criticality software while the neural network software was verified to NASA Class B standards. A floating limiter in the baseline controller software served as a monitor to the neural network outputs and allowed the Class B software to interact safely with the Class A baseline controller. Flight testing of the system was performed at NASA Dryden Flight Research Center, the results of which are quite promising. The IFCS program has made important steps in the demonstration of adaptive neural network technologies in high assurance systems such as piloted aircraft.

References

1. Hageman, J., Smith, M., Stachowiak, S.: Integration of Online Parameter Identification and Neural Network for In-flight Adaptive Control. Technical Report TR-2003-212028, NASA (2003)
2. Williams-Hayes, P.S.: Selected Flight Test Results for Online Learning Neural Network-Based Flight Control System. Technical Report TR-2004-212857 (2004)
3. Burken, J., Williams-Hayes, P.S., Kaneshige, J., Stachowiak, S.: Reconfigurable Control with Neural Network Augmentation for a Modified F-15 Aircraft. Technical Report TR-2006-213678, NASA (2006)
4. Soares, F., Burken, J.: A Flight Test Demonstration of On-line Neural Network Applications in Advanced Aircraft Flight Control System. In: International Conference on Computational Intelligence for Modeling Control and Automation and International Conference on Intelligent Agents Web Technologies and International Commerce (CIMCA), p. 136 (2006)
5. Dryden Flight Research Center Policy. Dryden Centerwide Procedure. Code S, Software Assurance, DCP-S-007, Revision B
6. Smolka, J.W., Walker, L.A., Major Johnson, G.H., Schkolnik, G.S., Berger, C.W., Conners, T.R., Orme, J.S., Shy, K.S., Wood, C.B.: F-15 ACTIVE Flight Research Program. In: 1996 Report to the Aerospace Profession Fortieth Symposium Proceedings, pp. 112–145 (1996)
7. Kim, B.S., Calise, A.J.: Nonlinear Flight Control Using Neural Networks. Journal of Guidance, Control, and Dynamics 20(1), 26–33 (1997)
8. Lavretsky, E.: Adaptive Control: Introduction, Overview and Applications. In: Lecture notes from IEEE Robust and Adaptive Control Workshop, May 20, 2005 (2008), `https://www.cds.caltech.edu/help/cms.php?op=wiki&wiki_op=view&id=140`
9. Flying Qualities of Piloted Vehicles. U.S. Department of Defense, MIL-STD-1797A (January 1990)
10. Slotine, J., Li, W.: Applied Nonlinear Control. Prentice-Hall, Englewood Cliffs (1991)
11. Cao, C., Hovakimyan, N., Lavretsky, E.: Application of L1 Adaptive Controller to Wing Rock. In: Proc. of AIAA Guidance, Navigation, and Controls Conference (2006)
12. Software Considerations in Airborne Systems and Equipment Certification. Document No. RTCA (Requirements and Technical Concepts for Aviation) /DO-178B (December 1992)
13. Bomben, C.R., Smolka, J.W., Bosworth, J.T., Williams-Hayes, P.S., Burken, J., Larson, R.R., Buschbacher, M.J., Maliska, H.A.: Development and Flight Testing of a Neural Network Based Flight Control System on the NF-15B Aircraft. In: Report to the Aerospace Profession Fiftieth Symposium Proceedings, pp. 214–240 (2006)
14. Bosworth, J.T., Williams-Hayes, P.S.: Flight Test Results from the NF-15B Intelligent Flight Control System (IFCS) Project with Adaptation to a Simulated Stabilator Failure. Technical Report TR-2007-214629, NASA (2007)

Stability, Convergence, and Verification and Validation Challenges of Neural Net Adaptive Flight Control

Nhan T. Nguyen and Stephen A. Jacklin

Abstract. This paper provides a discussion of challenges of neural net adaptive flight control and an examination of stability and convergence issues of adaptive control algorithms. Understanding stability and convergence issues with adaptive control is important in order to advance adaptive control to a higher technology readiness level. The stability and convergence of neural net learning law are investigated. The effect of unmodeled dynamics on learning law is examined. Potential improvements in the learning law and adaptive control architecture based on optimal estimation are presented. The paper provides a brief summary of the future research of the Integrated Resilient Aircraft Control (IRAC) in the area of adaptive flight control. The paper also discusses challenges and future research in verification and validation.

1 Introduction

While air travel remains the safest mode of transportation, accidents do occur on rare occasions with catastrophic consequences (1; 2). For this reason, the Aviation Safety Program under the Aeronautics Research Mission Directorate (ARMD) at NASA has created the Integrated Resilient Aircraft Control (IRAC) research project to advance the state of aircraft flight control and to provide onboard control resilience for ensuring safe flight in the presence of adverse conditions such as faults, damage, and/or upsets (3). These hazardous flight conditions can impose heavy demands on aircraft flight control systems in their abilities to enable a pilot to stabilize and navigate an aircraft safely. The flight control research to be developed by the IRAC project will involve many different disciplines including aerodynamics, aircraft flight dynamics, engine dynamics, airframe structural dynamics, and others. During off-nominal flight conditions, all these conditions can couple together to potentially overwhelm a pilot's ability to control an aircraft. The

J. Schumann, Y. Liu (Eds.): Appl. of Neural Networks in High Assur. Sys., SCI 268, pp. 77–110.
springerlink.com

IRAC research goal is to develop vehicle-centric multidisciplinary adaptive flight control approaches that can effectively deal with these coupled effects.

Modern aircraft are equipped with flight control systems that have been rigorously field-tested and certified by the Federal Aviation Administration. Functionally, an aircraft flight control system may be decomposed into an inner-loop and an outer-loop hierarchy. The outer-loop flight control is responsible for the aircraft guidance and navigation. It generates flight path or trajectory of the aircraft position based on pilot's inputs to the Flight Management System (FMS). The FMS provides pilots with capabilities for performing pre-flight planning, navigation, guidance, and performance management using built-in trajectory optimization tools for achieving operational efficiencies or mission objectives. The inner-loop flight control has a dual responsibility in tracking the trajectory commands generated by the outer-loop flight control and, more importantly, in stabilizing the aircraft attitude in the pitch, roll, and yaw axes. Because the aircraft must be stabilized rapidly in an off-nominal event such as upsets or damage, the inner-loop flight control must have a faster response time than the outer-loop flight control.

In a damage scenario, some part of an aerodynamic surface may become separated and, as a result, may cause an aircraft's center of gravity (C.G.) to shift unexpectedly (4). Furthermore, changes in aerodynamic characteristics can render a damaged aircraft unstable. Consequently, these effects can lead to a non-equilibrium flight that can adversely affect the ability of a flight control system to maintain aircraft stability. In other instances, reduced structural rigidity of a damaged airframe may manifest in elastic motions that can interfere with a flight control system, and potentially can result in excessive structural loading on critical lifting surfaces. Thus, in a highly dynamic, off-nominal flight environment with many sources of uncertainty, a flight control system must be able to cope with complex and uncertain aircraft dynamics.

The goal of the IRAC project is to arrive at a set of validated multidisciplinary integrated aircraft control design tools and techniques for enabling safe flight in the presence of adverse conditions (3). Aircraft stability and maneuverability in off-nominal flight conditions are critical to aircraft survivability. Adaptive flight control is identified as a technology that can improve aircraft stability and maneuverability. Stability of adaptive control remains a major challenge that prevents adaptive control from being implemented in human-rated or mission-critical flight vehicles (5; 6). Understanding stability issues with adaptive control, hence, will be important in order to advance adaptive control technologies. Thus, one of the objectives of IRAC adaptive control research is to develop metrics for assessing stability of adaptive flight control by extending the robust control concept of phase and gain margins to adaptive control. Furthermore, stability of adaptive control and the learning algorithms will be examined in the presence of unmodeled dynamics and exogenous disturbances. Another objective of the IRAC research is to advance adaptive control technologies that can better manage constraints placed on the aircraft. These constraints are dictated by limitations of actuator dynamics, aircraft structural load limits, frequency bandwidth, system latency, and others. New concepts of adaptive control will be developed to address these constraints.

2 Convergence and Stability of Neural Net Direct Adaptive Flight Control

Over the past several years, various adaptive control techniques have been investigated (7; 6; 8; 9; 10; 11; 12; 13; 14). Adaptive flight control provides a possibility for maintaining aircraft stability and performance by means of enabling a flight control system to adapt to system uncertainties. Meanwhile, a large area of research in intelligent control has emerged and spans many different applications such as spacecraft and aircraft flight control. One of the elements of an intelligent flight control is a neural network or any other machine learning algorithm designed to accommodate changes in aircraft dynamics due to system uncertainties. Neural network is known to be a good universal approximator of many nonlinear functions that can be used to model system uncertainties (15; 16; 17). In the implementation of an intelligent flight control, the neural network is usually incorporated within a direct adaptive control architecture to provide an augmentation to a pilot command. The neural network estimates the system uncertainties and outputs directly a command augmentation signal that compensates for changes in aircraft dynamics.

Research in adaptive control has spanned several decades, but stability robustness in the presence of unmodeled dynamics, parameter uncertainties, or disturbances as well as the issues with verification and validation of adaptive flight control software still pose as major challenges (6; 5). Adaptive control laws may be divided into direct and indirect approaches. Indirect adaptive control methods typically compute control parameters from on-line learning neural networks that perform plant parameter estimation (24). Parameter identification techniques such as recursive least-squares and neural networks have been used in indirect adaptive control methods (8). In recent years, model-reference direct adaptive control using neural networks has been a topic of great research interest (9; 10; 11). Lyapunov stability theory has been used to test theoretical stability of neural network learning laws (9).

NASA has been developing an intelligent flight control technology based on the work by Rysdyk and Calise (9). An architecture of the intelligent flight control is as shown in Fig. 1. Recently, this technology has been demonstrated on an F-15 fighter aircraft (18). The intelligent flight control uses a model-reference, direct adaptive, dynamic inversion control approach. The neural network direct adaption is designed to provide consistent handling qualities without requiring extensive gain-scheduling or explicit system identification. This particular architecture uses both pre-trained and on-line learning neural networks and a reference model to specify desired handling qualities. Pre-trained neural networks are used to provide estimates of aerodynamic stability and control characteristics. On-line learning neural networks are used to compensate for errors and adapt to changes in aircraft dynamics. As a result, consistent handling qualities may be achieved across flight conditions.

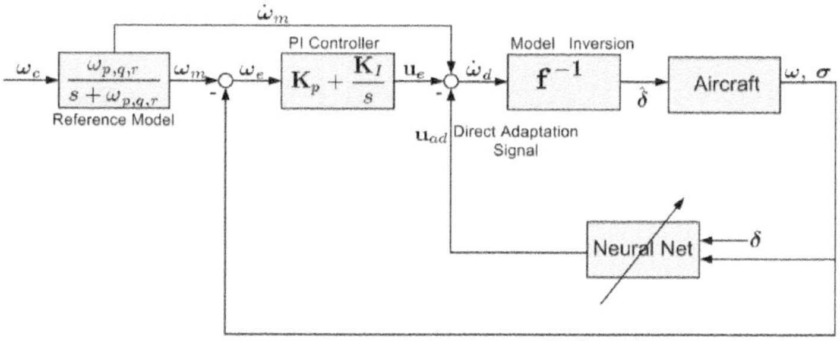

Fig. 1 Neural Net Adaptive Flight Control Architecture

Recent flight test results demonstrate the potential benefits of adaptive control technology in improving aircraft flight control systems in the presence of adverse flight conditions due to failures (19). The flight test results also point out the needs for further research to increase the understanding of effectiveness and limitations of the direct adaptive flight control.

One potential problem with the neural net direct adaptive flight control is that high gain control used to facilitate aggressive learning to reduce the error signal rapidly can potentially result in a control augmentation command that may saturate the control authority or excite unmodeled dynamics of the plant that can adversely affect the stability of the direct adaptive learning law.

2.1 Direct Adaptive Control Approach

Adaptive flight control is designed to accommodate parametric and system uncertainties in the presence of adverse conditions such as upset, failures, and damage (4). The true aircraft dynamics may be described by a linear model about its trim point in a flight envelope

$$\dot{\boldsymbol{\omega}} = \dot{\boldsymbol{\omega}}^* + \Delta\dot{\boldsymbol{\omega}} = \mathbf{F}_1\boldsymbol{\omega} + \mathbf{F}_2\boldsymbol{\sigma} + \mathbf{G}\boldsymbol{\delta} \tag{1}$$

where $\boldsymbol{\omega} = \begin{bmatrix} p \; q \; r \end{bmatrix}^\top$ is the aircraft angular rate, $\boldsymbol{\sigma} = \begin{bmatrix} \Delta\alpha \; \Delta\beta \; \Delta\phi \end{bmatrix}^\top$ is an incremental trim state vector to maintain trim condition, $\Delta\dot{\boldsymbol{\omega}}$ is the unknown aircraft dynamics due to parametric uncertainties, and $\dot{\boldsymbol{\omega}}^*$ is the nominal aircraft dynamics described by

$$\dot{\boldsymbol{\omega}}^* = \mathbf{F}_1^*\boldsymbol{\omega} + \mathbf{F}_2^*\boldsymbol{\sigma} + \mathbf{G}^*\boldsymbol{\delta} \tag{2}$$

where $\mathbf{F}_1^*$, $\mathbf{F}_2^*$, and $\mathbf{G}^*$ as the nominal plant matrices which are assumed to be known (4).

The input to the dynamic inversion controller is a desired acceleration command

$$\dot{\omega}_d = \mathbf{F}_1^* \omega + \mathbf{F}_2^* \sigma + \mathbf{G}^* \delta \tag{3}$$

which is computed from the reference model acceleration, a proportional-integral feedback control, and an adaptive signal

$$\dot{\boldsymbol{\omega}}_d = \dot{\boldsymbol{\omega}}_m + \mathbf{K}_p \boldsymbol{\omega}_e + \mathbf{K}_i \int_0^t \boldsymbol{\omega}_e d\tau - \mathbf{u}_{ad} \tag{4}$$

where $\boldsymbol{\omega}_e = \boldsymbol{\omega}_m - \boldsymbol{\omega}$ is defined as the rate error signal which is the difference between a reference model rate $\boldsymbol{\omega}_m$ and the actual aircraft rate, $\mathbf{u}_{ad}$ is the neural net direct adaptive signal, and $\dot{\boldsymbol{\omega}}_m$ is the acceleration of a first-order reference model which filters a pilot command into a reference acceleration command that provides a suitable handling characteristic.

The difference in the aircraft dynamics gives rise to the modeling error $\boldsymbol{\varepsilon}$ which is defined as

$$\boldsymbol{\varepsilon} = \dot{\boldsymbol{\omega}} - \dot{\boldsymbol{\omega}}_d = \Delta \mathbf{F}_1 \boldsymbol{\omega} + \Delta \mathbf{F}_2 \boldsymbol{\sigma} + \Delta \mathbf{G} \boldsymbol{\delta} \tag{5}$$

The modeling error that exists in the controller causes the aircraft states to deviate from its desired states. This results in a tracking error dynamics which are described by

$$\dot{\mathbf{e}} = \mathbf{A}\mathbf{e} + \mathbf{B}\left(\mathbf{u}_{ad} - \boldsymbol{\varepsilon}\right) \tag{6}$$

where $\mathbf{e} = \left[\int_0^t \boldsymbol{\omega}_e d\tau \ \boldsymbol{\omega}_e \right]^\top$ is a proportional-integral tracking error, and $\mathbf{A}$ and $\mathbf{B}$ are matrices defined as

$$\mathbf{A} = \begin{bmatrix} \mathbf{0} & \mathbf{I} \\ -\mathbf{K}_i & -\mathbf{K}_p \end{bmatrix}, \ \mathbf{B} = \begin{bmatrix} \mathbf{0} \\ \mathbf{I} \end{bmatrix} \tag{7}$$

where $\mathbf{K}_i = \text{diag}\left(k_{i,1}, k_{i,2}, k_{i,3}\right)$ and $\mathbf{K}_p = \text{diag}\left(k_{p,1}, k_{p,2}, k_{p,3}\right)$ are diagonal matrices of integral and proportional gains corresponding to aircraft angular rates p, q, and r.

To compensate for the parametric uncertainties, the neural net direct adaptive signal $\mathbf{u}_{ad}$ uses a linear-in-parameter sigma-pi network proposed by Rysdyk and Calise (9). This network is further modified (20) to include additional inputs as shown in Fig. 2. Associated with the neural network is a set of weights that are adjusted on-line in order to enable the flight control system to adapt to changes in aircraft dynamics. The neural net direct adaptive signal and the weights are computed by

$$\mathbf{u}_{ad} = \mathbf{W}^\top \boldsymbol{\beta} \tag{8}$$

$$\dot{\mathbf{W}} = -\Gamma \left(\boldsymbol{\beta} \mathbf{e}^\top \mathbf{P} \mathbf{B} + \mu \left\| \mathbf{e}^\top \mathbf{P} \mathbf{B} \right\| \mathbf{W} \right) \tag{9}$$

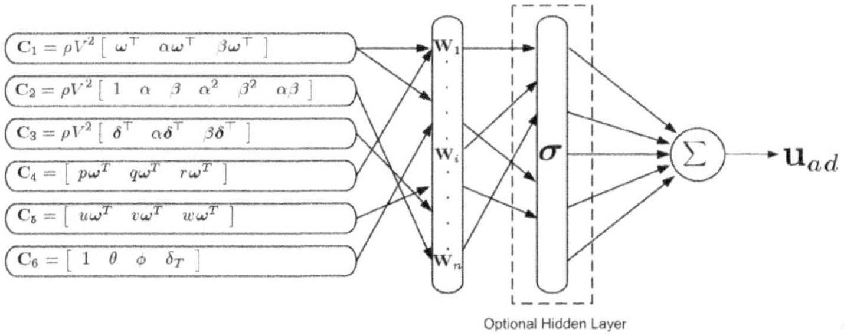

Fig. 2 Sigma-Pi Neural Network

where Γ is the learning rate, μ is the e-modification term, $\boldsymbol{\beta}$ is a basis function comprising neural net inputs $\mathbf{C}_i$, $i = 1, \ldots, 6$, and $\mathbf{P}$ is the solution of the Lyapunov equation

$$\mathbf{A}^\top \mathbf{P} + \mathbf{P}^\top \mathbf{A} = -\mathbf{Q} \tag{10}$$

where $\mathbf{Q}$ is a positive-definite matrix.

if $\mathbf{Q} = c\mathbf{I}$ where $c > 0$, then $\mathbf{P}$ is evaluated as

$$\mathbf{P} = \begin{bmatrix} \mathbf{P}_{11} & \mathbf{P}_{12} \\ \mathbf{P}_{12} & \mathbf{P}_{22} \end{bmatrix} = \frac{c}{2} \begin{bmatrix} \mathbf{K}_i^{-1}\mathbf{K}_p + \mathbf{K}_p^{-1}(\mathbf{K}_i + \mathbf{I}) & \mathbf{K}_i^{-1} \\ \mathbf{K}_i^{-1} & \mathbf{K}_p^{-1}\left(\mathbf{I} + \mathbf{K}_i^{-1}\right) \end{bmatrix} > \mathbf{0} \tag{11}$$

2.2 Stability and Convergence

A key challenge with neural net adaptive flight control is to make the learning algorithm sufficiently robust. Robustness relates to the stability and convergence of the learning algorithm. Stability is a fundamental requirement of any dynamical system that ensures a small disturbance would not grow to a large deviation from an equilibrium. For systems with high assurance such as human-rated or mission-critical flight vehicles, stability of adaptive systems is of paramount importance. Without guaranteed stability, such adaptive control algorithms cannot be certified for operation in high-assurance systems. Unfortunately, the stability of adaptive controllers in general and neural net adaptive controllers in particular remains unresolved. The notion of a self-modifying flight control law using an artificial neural net learning process whose outputs may be deemed as non-deterministic is a major huddle to overcome.

Another criterion for robustness is the convergence of the neural net learning algorithm. Neural networks are used as universal nonlinear function approximators. In the case of the adaptive flight control, the networks approximate the unknown modeling error that is used to adjust effectively the control gains to maintain a desired handling quality. Convergence requires stability and a proper design of the weight update law. It is conceivable that even though a learning algorithm is stable, the neural net weights may not converge to correct values. Thus, accurate convergence is also important since this is directly related to the flight control performance.

Referring to Eq. (6), if the direct adaptive signal $\mathbf{u}_{ad}$ could perfectly cancel out the modeling error ε, then the tracking error would tend to zero asymptotically. Therefore, ideally the desired acceleration $\dot{\omega}_d$ would perfectly track the reference model acceleration $\dot{\omega}_m$. In practice, there is always some residual modeling error in the adaptation, so asymptotic stability of the tracking error is not guaranteed. Instead, a weaker uniform stability of the tracking error can be established by the Lyapunov stability theory. The tracking error dynamics is then bounded from below by

$$\dot{\mathbf{e}} \leq \mathbf{A}\mathbf{e} + \mathbf{B}\tilde{\mathbf{W}}^{\top}\boldsymbol{\beta} + \boldsymbol{\Delta}_1 \tag{12}$$

where $\tilde{\mathbf{W}}$ is the neural net weight matrix variation and $\boldsymbol{\Delta}_1$ is the approximation error upper bound defined as

$$\boldsymbol{\Delta}_1 = -\mathbf{A}\left|\mathbf{A}^{-1}\sup_{\boldsymbol{\beta}}\left|\mathbf{B}\left(\mathbf{W}^{*\top}\boldsymbol{\beta} - \varepsilon\right)\right|\right| \tag{13}$$

where $\mathbf{W}^*$ is the ideal weight matrix which is equal to

$$\mathbf{W}^* = \mathbf{W} - \tilde{\mathbf{W}} \tag{14}$$

It can be shown by the Lyapunov analysis (20) that there exists a δ-neighborhood that the tracking error remains bounded from below

$$\delta = \inf_{\omega}\|\mathbf{e}\| = \frac{\lambda_{max}(\mathbf{P})\|\mathbf{B}\|}{2\lambda_{min}(\mathbf{Q})}\left(4\|\boldsymbol{\Delta}_1\| + \mu\|\mathbf{W}^*\|^2\right) \tag{15}$$

where λ_{max} and λ_{min} denote the maximum and minimum eigenvalues, respectively.

Thus, if δ is sufficiently small, then the tracking error will be reasonably close to zero. This means that the neural net approximation error is also small and the tracking performance is good.

The neural net weight update law in Eq. (9) is nonlinear due to the product term of the basis function $\boldsymbol{\beta}$ and the tracking error $\mathbf{e}$. Stability of nonlinear systems is usually analyzed by the Lyapunov method. However, the concept of phase and gain margin for linear systems cannot be extended to nonlinear adaptive control. The linear control margin concept can provide understanding stability margin of adaptive control that enables a more robust adaptive learning law to be synthesized. This is only possible if the neural net weight update law is linearized at a certain point in time with the neural net weights held constant. As adaptation occurs, the neural net weights vary with time. Hence, the time at which to freeze the neural net weights (for calculation) must correspond to a worst-case stability margin. This can be a challenge. Therefore, this paper introduces a new method for analyzing stability and convergence of the nonlinear neural net weight update law using an error bound analysis, which enables the dominant linear component of the nonlinear neural net weight update law to be approximated from Eq. (9) in a small time window $t \in (t_0 - T, t_0]$ without linearizing the adaptive control law at an instance in time (21; 22). Towards that end, the time derivative of the variation in the neural net adaptive signal is bounded by (21)

$$\dot{\boldsymbol{\psi}} \leq -\Gamma \left(\alpha \mathbf{B}^\top \mathbf{P} \mathbf{e} + \mu \gamma \boldsymbol{\psi}\right) + \boldsymbol{\Delta}_2 \tag{16}$$

if there exists a constant $\alpha > 0$ which is defined as a persistent excitation (PE) parameter such that

$$\|\boldsymbol{\beta}\|^2 = \frac{1}{T} \int_{t_0 - T}^{t_0} \boldsymbol{\beta}^\top \boldsymbol{\beta} d\tau \geq \alpha \tag{17}$$

and a constant γ such that

$$\gamma = \inf_{\omega \forall t \in (t_0 - T, t_0]} \left\| \mathbf{e}^\top \mathbf{P} \mathbf{B} \right\| \tag{18}$$

where $\boldsymbol{\psi} = \tilde{\mathbf{W}}^\top \boldsymbol{\beta}$ is the variation in the neural net adaptive signal and $\boldsymbol{\Delta}_2$ is a constant upper bound

$$\boldsymbol{\Delta}_2 = \sup_{\boldsymbol{\beta} \forall t \in (t_0 - T, t_0]} \left| \tilde{\mathbf{W}}^\top \dot{\boldsymbol{\beta}} - \Gamma \mu \gamma \mathbf{W}^{*\top} \boldsymbol{\beta} \right| \tag{19}$$

If the error bound is small, then the approximate linear behavior of the weight update law becomes dominant in the small time window $t \in (t_0 - T, t_0]$. Therefore, this enables the stability and convergence to be analyzed approximately in a linear time invariant sense using the following equation

$$\begin{bmatrix} \dot{\mathbf{e}} \\ \dot{\boldsymbol{\psi}} \end{bmatrix} \leq \begin{bmatrix} \mathbf{A} & \mathbf{B} \\ -\Gamma \alpha \mathbf{B}^\top \mathbf{P} & -\Gamma \mu \gamma \mathbf{I} \end{bmatrix} \begin{bmatrix} \mathbf{e} \\ \boldsymbol{\psi} \end{bmatrix} + \begin{bmatrix} \boldsymbol{\Delta}_1 \\ \boldsymbol{\Delta}_2 \end{bmatrix} \tag{20}$$

If $\Re(\lambda) < 0$, then the rate of convergence is established as

$$\begin{bmatrix} \mathbf{e} \\ \boldsymbol{\psi} \end{bmatrix} \leq \boldsymbol{\Phi} e^{\boldsymbol{\Lambda} t} \boldsymbol{\Phi}^{-1} \begin{bmatrix} \mathbf{e}(0) \\ \boldsymbol{\psi}(0) \end{bmatrix} - \left(\mathbf{I} - \boldsymbol{\Phi} e^{\boldsymbol{\Lambda} t} \boldsymbol{\Phi}^{-1}\right) \begin{bmatrix} \mathbf{A} & \mathbf{B} \\ -\Gamma \alpha \mathbf{B}^\top \mathbf{P} & -\Gamma \mu \gamma \mathbf{I} \end{bmatrix}^{-1} \begin{bmatrix} \boldsymbol{\Delta}_1 \\ \boldsymbol{\Delta}_2 \end{bmatrix} \tag{21}$$

where $\boldsymbol{\Lambda} = \operatorname{diag}(\lambda_1, \ldots, \lambda_9)$ is a diagonal eigenvalue matrix and $\boldsymbol{\Phi}$ is a matrix of right eigenvectors of the leading matrix in Eq. (20).

The equilibrium is therefore uniformly asymptotically stable and converges to

$$\lim_{t_0 \to \infty} \sup_{t \in (t_0 - T, t_0]} \left\| \begin{bmatrix} \mathbf{e} \\ \boldsymbol{\psi} \end{bmatrix} \right\| = \left\| \begin{bmatrix} \mathbf{A} & \mathbf{B} \\ -\Gamma \alpha \mathbf{B}^\top \mathbf{P} & -\Gamma \mu \gamma \mathbf{I} \end{bmatrix}^{-1} \begin{bmatrix} \boldsymbol{\Delta}_1 \\ \boldsymbol{\Delta}_2 \end{bmatrix} \right\| \tag{22}$$

By Holder's inequality, the convergence radius can be expressed as

$$\lim_{t\to\infty} \sup_{t\in(t_0-T,t_0]} \left\| \begin{bmatrix} \mathbf{e} \\ \boldsymbol{\psi} \end{bmatrix} \right\| \leq \rho\left(\boldsymbol{\Lambda}^{-1}\right) \left\| \begin{bmatrix} \boldsymbol{\Delta}_1 \\ \boldsymbol{\Delta}_2 \end{bmatrix} \right\| \tag{23}$$

Thus, $\boldsymbol{\Delta}_1$ and $\boldsymbol{\Delta}_2$ should be kept as small as possible for the tracking error and the neural net weight matrix variation to converge as close to zero as possible.

The stability of the learning law requires the leading matrix to be negative definite. This can be established by computing its eigenvalues from the following characteristic equation

$$(s + \Gamma\mu\gamma)\det\left(s\mathbf{I} - \mathbf{A} + \frac{\Gamma\alpha\mathbf{B}\mathbf{B}^\top\mathbf{P}}{s + \Gamma\mu\gamma}\right) = 0 \tag{24}$$

Upon expansion of Eq. (24), we get

$$\prod_{j=1}^{3}\Big[s^3 + (k_{p,j} + \Gamma\mu\gamma)\, s^2 + (k_{i,j} + \Gamma\alpha_0 p_{22,j} + \Gamma\mu\gamma k_{p,j})\, s + \Gamma\alpha_0 p_{12,j} + \Gamma\mu\gamma k_{i,j}\Big] = 0 \tag{25}$$

where $p_{12,j}$ and $p_{22,j}$ are the $j-$th diagonal elements of $\mathbf{P}_{12}$ and $\mathbf{P}_{22}$, respectively.

To estimate the eigenvalues, the characteristic equation is factored as

$$(s + \Gamma a_j)\Big[s^2 + (k_{p,j} + \Gamma\mu\gamma - \Gamma a_j)\, s + k_{i,j} + \Gamma\mu\gamma k_{p,j} + \Gamma\alpha_0 p_{22,j} - \Gamma a_j\,(k_{p,j} + \Gamma\mu\gamma - \Gamma a_j)\Big] + r_j = 0 \tag{26}$$

where a and the residue r are defined as

$$a_j = (k_{i,j} + \Gamma\mu\gamma k_{p,j} + \Gamma\alpha_0 p_{22,j})^{-1}\,(\mu\gamma k_{i,j} + \alpha_0 p_{12,j}) \tag{27}$$

$$r_j = \Gamma\,(\mu\gamma k_{i,j} + \alpha_0 p_{12,j}) - \Gamma a_j\Big[k_{i,j} + \Gamma\,(\mu\gamma k_{p,j} + \alpha_0 p_{22,j}) - \Gamma a_j\,(k_{p,j} + \Gamma\mu\gamma - \Gamma a_j)\Big] \tag{28}$$

In order for the neural net weights to converge, stability of the tracking error and neural net weight update law must be established. Consider the following cases:

1. $\underline{\Gamma \ll 1}$:
   This corresponds to a small learning rate or slow adaptation. We see that

$$a_j \approx \mu\gamma + \alpha_0 k_{i,j}^{-1} p_{12,j} \tag{29}$$

$$r \approx 0 \tag{30}$$

Neglecting second-order terms of Γ, The approximate eigenvalues are then found to be

$$s_{1,2,j} \approx -\frac{k_{p,j} - \Gamma\alpha_0 k_{i,j}^{-1} p_{12,j}}{2} \pm j\left[k_{i,j} - \frac{k_{p,j}^2}{4} + \Gamma\alpha_0\left(p_{22,j} - \frac{k_{p,j}k_{i,j}^{-1}p_{12,j}}{2}\right)\right]^{\frac{1}{2}} \tag{31}$$

$$s_{3,j} \approx -\Gamma\left(\mu\gamma + \alpha_0 k_{i,j}^{-1} p_{12,j}\right) \tag{32}$$

if $k_{i,j} > k_{p,j}^2/4$ and $p_{22,j} > k_{p,j}k_{i,j}^{-1}p_{12,j}/2$.

The eigenvalue $s_{3,j}$ is proportional to the learning rate and thus indicates that the neural net learning is very slow if Γ is small. The convergence radius is then approximately equal to

$$\lim_{t\to\infty} \sup_{t\in(t_0-T,t_0]} \left\|\begin{bmatrix}\mathbf{e}\\ \boldsymbol{\psi}\end{bmatrix}\right\| \le \frac{1}{\Gamma\left(\mu\gamma + \alpha_0 k_{i,j}^{-1}p_{12,j}\right)} \left\|\begin{bmatrix}\boldsymbol{\Delta}_1\\ \boldsymbol{\Delta}_2\end{bmatrix}\right\| \tag{33}$$

Since Γ is small, the convergence radius is large and thus the neural network would not achieve a good convergence. The convergence also depends on the persistent excitation parameter α and the e-modification parameter μ. If $\mu = 0$, the convergence of the neural net weights requires a persistent excitation.

2. $\underline{\Gamma \gg 1}$:
This corresponds to a large learning rate or fast adaptation. For $\mu = 0$, we obtain

$$\Gamma a_j \approx p_{22,j}^{-1} p_{12,j} \tag{34}$$

$$r \approx -\Gamma a_j\left[k_{i,j} - \Gamma a_j\left(k_{p,j} - \Gamma a_j\right)\right] \tag{35}$$

Since Γa is finitely small even though Γ is large, then the residue r is also finitely small compared to s which is large. The approximate eigenvalues are obtained as

$$s_{1,2,j} \approx -\frac{k_{p,j} - p_{22,j}^{-1}p_{12,j}}{2} \pm j\left[k_{i,j} - \frac{k_{p,j}^2}{4} - \frac{p_{22,j}^{-1}p_{12,j}k_{p,j}}{2} + \frac{3p_{22,j}^{-2}p_{12,j}^2}{4} + \Gamma\alpha_0 p_{22,j}\right]^{\frac{1}{2}} \tag{36}$$

$$s_{3,j} \approx -p_{22,j}^{-1}p_{12,j} \tag{37}$$

The convergence radius is then approximately equal to

$$\lim_{t\to\infty} \sup_{t\in(t_0-T,t_0]} \left\| \begin{bmatrix} \mathbf{e} \\ \boldsymbol{\psi} \end{bmatrix} \right\| \leq p_{22,j} p_{12,j}^{-1} \left\| \begin{bmatrix} \boldsymbol{\Delta}_1 \\ \boldsymbol{\Delta}_2 \end{bmatrix} \right\| \tag{38}$$

If μ is sufficiently large, then in the limit for a very large learning rate, the neural net weights in theory would converge to

$$\lim_{\Gamma\to\infty} \boldsymbol{\psi} = -\frac{\alpha \mathbf{B}^\top \mathbf{P} \mathbf{e}}{\mu\gamma} \tag{39}$$

and the tracking error dynamics become

$$\dot{\mathbf{e}} \leq \left(\mathbf{A} - \frac{\alpha \mathbf{B}\mathbf{B}^\top \mathbf{P}}{\mu\gamma} \right) \mathbf{e} + \boldsymbol{\Delta}_1 \tag{40}$$

Thus, in theory, a large learning rate would improve stability and convergence. However, in practice there exists a potential problem with a large learning rate. Equation (36) shows that increasing the learning rate to a large value with $\mu = 0$ will result in a high frequency oscillation in the adaptive signal (23). This high frequency oscillation can result in excitation of unmodeled dynamics that may be present in the system and therefore can lead to a possibility of instability since the effects of unmodeled dynamics are not accounted in the Lyapunov analysis of the neural net weight update law (24). When μ is large, increasing the learning rate does not lead to a high frequency oscillation which tends to improve robustness, but the convergence is poorer. Thus, high-gain learning should be avoided in adaptive control.

Another source of instability is measurement noise which will also influence high-gain controllers to make over-corrections. Stated loosely, the neural network then maps the spurious noise signals to small changes in the control commands, leading to a loss of true learning and instability. For this reason, learning is typically disabled when the error falls below a certain value by applying a dead band.

To illustrate the effects of learning rate, a simulation was performed for a damaged twin-engine generic transport model (GTM) (25), as shown in Fig. 3. A wing damage simulation was performed with a 25% left wing separation. The neural net direct adaptive control is implemented to maintain tracking performance of the damaged aircraft. A pitch doublet maneuver is commanded while the roll and yaw rates are regulated near zero.

Fig. 4 illustrates the effect of learning rate without the e-modification term, i.e, $\mu = 0$. Without adaptation, the performance of the flight control is very poor as significant overshoots occur. With adaptation, good tracking performance can be obtained. As the learning rate increases, the tracking error becomes smaller but high frequency signals also appear that are consistent with the analysis above.

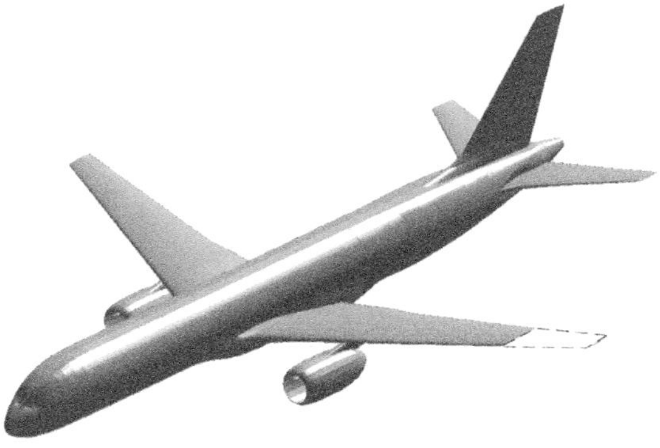

Fig. 3 Generic Transport Mode

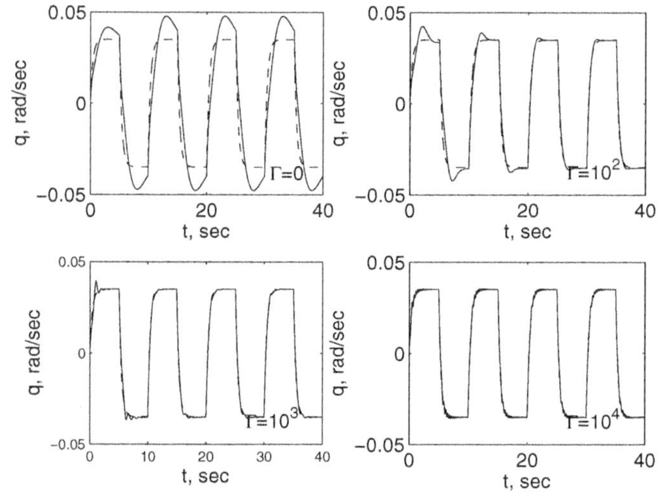

Fig. 4 Pitch Rate ($\mu = 0$)

Fig. 5 is a plot of selected neural net weights for various learning rates. As can be seen, large learning rate causes high frequency oscillations in the weights. The convergence of the neural net weights $W_{q,q}$ and W_{q,δ_e} associated with linear elements q and δ_e for the pitch rate are poor. Neither of these weights would actually converge to their correct values. Thus, convergence accuracy is not demonstrated.

Fig. 6 illustrates the effect of the e-modification parameter μ. As μ increases, the high-frequency amplitude reduces but the tracking error becomes worse. Eventually, with large enough value of μ, the learning essentially ceases. Fig. 7 is the plot of selected neural net weights. Clearly, with increasing μ, the weights are driven to zero, thereby reducing the learning of the neural network. This is consistent with the analysis above which shows the

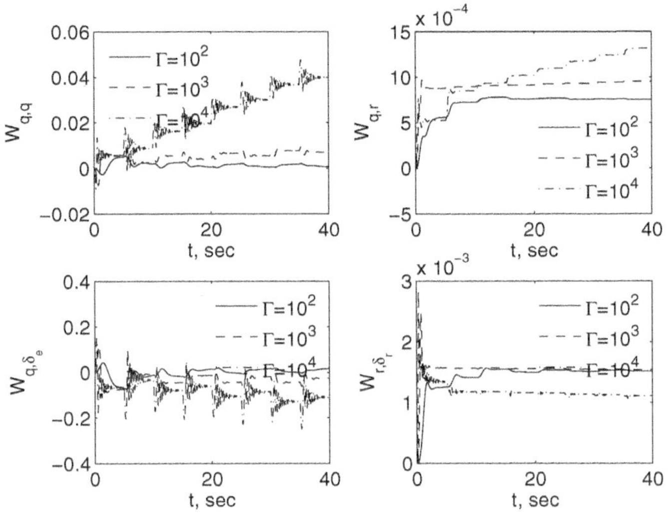

Fig. 5 Neural Net Weight Learning ($\mu = 0$)

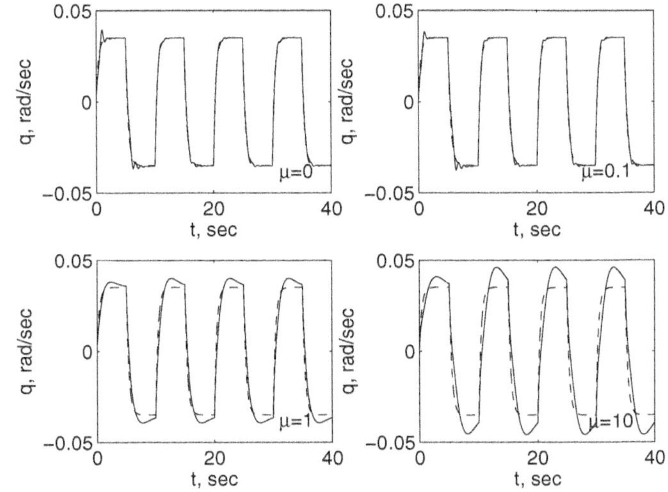

Fig. 6 Pitch Rate with $\mu \neq 0$ ($\Gamma = 1000$)

for learning rate with $\mu \neq 0$, the effect is essentially the same as that with small learning rate.

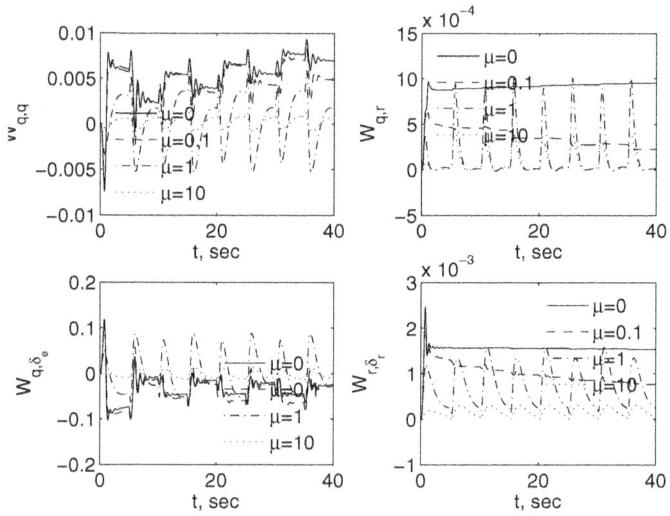

Fig. 7 Neural Net Weight Learning ($\Gamma = 1000$)

2.3 Unmodeled Dynamics

Unmodeled dynamics are secondary dynamics that are ignored in the system dynamics. Usually, the small effects of unmodeled dynamics can be quite important, but sometimes are not explicitly accounted for in a control design due to the complexity of the physical modeling. An example of secondary dynamics is the structural dynamics of the airframe undergoing elastic deformation in flight. Typically, if the controller possesses sufficient gain and phase margins, then the controller can be verified during validation to ensure that it is sufficiently robust to safeguard against potentially destabilizing effects of unmodeled dynamics.

Unmodeled dynamics can have a profound effect on the stability of the learning algorithms for adaptive control. Even though an adaptive control algorithm may demonstrate stability for a dominant system dynamics, it can become unstable when a small unmodeled dynamics is present in the system. This can be shown by considering the following example of aircraft dynamics

$$\dot{\boldsymbol{\omega}} = \mathbf{F}_1\boldsymbol{\omega} + \mathbf{F}_2\boldsymbol{\sigma} + \mathbf{G}\boldsymbol{\delta} - \mathbf{z} \tag{41}$$

where $\mathbf{z}$ is a small parasitic state that represents the effects of unmodeled dynamics that have a certain property which can be described by

$$\epsilon\dot{\mathbf{z}} = -\mathbf{z} - \eta\mathbf{G}^*\boldsymbol{\delta} \tag{42}$$

where $\epsilon > 0$ and $\eta > 0$ are small parameters. Additionally, let the measurement output of the aircraft be the angular rate vector

$$\mathbf{y} = \boldsymbol{\omega} \tag{43}$$

Since the dynamic inversion controller is

$$\boldsymbol{\delta} = \mathbf{G}^{*-1}\left(\dot{\boldsymbol{\omega}}_m - \mathbf{B}^\top \mathbf{A}\mathbf{e} - \mathbf{W}^\top \boldsymbol{\beta} - \mathbf{F}_1^* \boldsymbol{\omega} - \mathbf{F}_2^* \boldsymbol{\sigma}\right) \tag{44}$$

then, the unmodeled dynamics are expressed as

$$\epsilon \dot{\mathbf{z}} \leq -\mathbf{z} + \eta \mathbf{G}\mathbf{G}^{*-1}\left(\mathbf{B}^\top \mathbf{A}\mathbf{e} + \boldsymbol{\psi}\right) + \epsilon \boldsymbol{\Delta}_3 \tag{45}$$

where

$$\boldsymbol{\Delta}_3 = \frac{\eta}{\epsilon} \sup_{\boldsymbol{\omega}} \left| \mathbf{G}\mathbf{G}^{*-1}\left(\mathbf{W}^{*\top} \boldsymbol{\beta} + \mathbf{F}_1^* \boldsymbol{\omega} + \mathbf{F}_2^* \boldsymbol{\sigma} - \dot{\boldsymbol{\omega}}_m\right) \right| \tag{46}$$

The dynamics of the combined system are bounded by

$$\begin{bmatrix} \dot{\mathbf{e}} \\ \dot{\boldsymbol{\psi}} \\ \dot{\mathbf{z}} \end{bmatrix} \leq \begin{bmatrix} \mathbf{A} & \mathbf{B} & \mathbf{B} \\ -\Gamma \alpha \mathbf{B}^\top \mathbf{P} & -\Gamma \mu \gamma & \mathbf{0} \\ \frac{\eta}{\epsilon} \mathbf{G}\mathbf{G}^{*-1} \mathbf{B}^\top \mathbf{A} & \frac{\eta}{\epsilon} \mathbf{G}\mathbf{G}^{*-1} & -\frac{1}{\epsilon}\mathbf{I} \end{bmatrix} \begin{bmatrix} \mathbf{e} \\ \boldsymbol{\psi} \\ \mathbf{z} \end{bmatrix} + \begin{bmatrix} \boldsymbol{\Delta}_1 \\ \boldsymbol{\Delta}_2 \\ \boldsymbol{\Delta}_3 \end{bmatrix} \tag{47}$$

We consider the case when $\mu = 0$ and $\mathbf{G} = \mathbf{G}^*$. For the purpose of illustration, we consider only a proportional control so that $\mathbf{K}_i = \mathbf{0}$. Then $\mathbf{B} = \mathbf{I}$, and $\mathbf{B}^\top \mathbf{P} = \mathbf{P}_{22}$. The characteristic equation is evaluated as

$$\det \begin{bmatrix} s + \mathbf{K}_p & -\mathbf{I} & -\mathbf{I} \\ \Gamma \alpha \mathbf{P}_{22} & s\mathbf{I} & \mathbf{0} \\ \frac{\eta}{\epsilon}\mathbf{K}_p & -\frac{\eta}{\epsilon} & \left(s + \frac{1}{\epsilon}\right)\mathbf{I} \end{bmatrix} = \prod_{j=1}^{3} \left[s^3 + \left(k_{p,j} + \frac{1}{\epsilon}\right) s^2 + \left(\frac{k_{p,j}}{\epsilon} + \frac{\eta}{\epsilon} k_{p,j} + \Gamma \alpha p_{22,j}\right) s + \frac{\Gamma \alpha p_{22,j}}{\epsilon}(1 + \eta) \right] = 0 \tag{48}$$

Applying the Routh-Hurwitz criterion, the following inequality is required for stability of the system; i.e., the closed-loop poles are on the open left-half s-plane:

$$\left(k_{p,j} + \frac{1}{\epsilon}\right)\left(\Gamma \alpha p_{22,j} + \frac{1}{\epsilon} k_{p,j} + \frac{\eta}{\epsilon} k_{p,j}\right) - \frac{1}{\epsilon} \Gamma \alpha p_{22,j} - \frac{\eta}{\epsilon} \Gamma \alpha p_{22,j} > 0 \tag{49}$$

If $\epsilon \ll k_{p,j}^{-1}$, we can approximate the inequality by

$$\frac{1}{\epsilon}\left(\Gamma \alpha p_{22,j} + \frac{1}{\epsilon} k_{p,j} + \frac{\eta}{\epsilon} k_{p,j}\right) - \frac{1}{\epsilon} \Gamma \alpha p_{22,j} - \frac{\eta}{\epsilon} \Gamma \alpha p_{22,j} > 0 \tag{50}$$

This implies that there exists an upper limit of the learning rate

$$\Gamma < \frac{k_{p,j}\,(1+\eta)}{\epsilon\eta\alpha p_{22,j}} \tag{51}$$

Thus, if the learning rate is sufficient large and exceeds this upper bound, it is conceivable that stability requirements could be violated, thereby resulting in instability of the adaptive control algorithm. Clearly, large learning rate can lead to a fast adaptation and large parasitic state $\mathbf{z}$ which acts as a disturbance to the dominant system dynamics, thereby leading to faulty adaptation that can result in unbounded solutions. For stability and bounded solutions, the learning rate should be kept small such that the speed of adaptation should be slow relative to the speed of the parasitic state $\frac{1}{\epsilon}$ (24). However, small learning rate will result in less than desired command-tracking performance. Therefore, a suitable learning rate is one that strives to achieve a reasonable balance between stability and performance of an adaptive flight control. In essence, this means that the acceptable gains for a stable neural net learning law must be found by trial and error, yet guided by theory and known structures of unmodeled dynamics, if possible.

Fig. 8 illustrates the effects of unmodeled dynamics on the learning of the adaptive control. The unmodeled dynamics result in a poor response when adaptation is off, but once adaptation is switched on, improvements can be immediately obtained. Comparing with Fig. 4 , the high learning rate excites the unmodeled dynamics, thereby causing an increase in the undesired high frequency noise in the adaptive signals. With sufficiently large learning rate, the weight update law would become unstable.

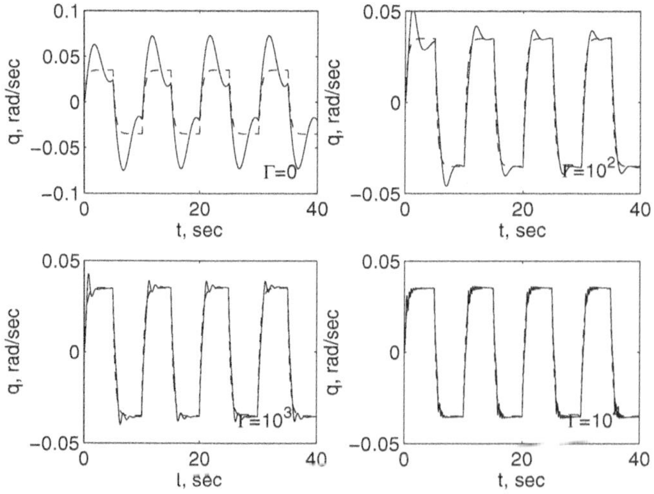

Fig. 8 Pitch Rate with Unmodeled Dynamics ($\mu = 0$, $\epsilon = 0.1$, $\eta = 0.1$)

3 Potential Improvements

In the neural net weight update law in Eq. (9), the tracking error signal $\mathbf{e}$ is used for adaptation. This adaptive law is based on the stability analysis of the tracking error using the Lyapunov method (9). Examining Eq. (6), one sees that if the adaptive signal $\mathbf{u}_{ad}$ could cancel out the modeling error ε, then the tracking error will tend to zero asymptotically. Thus, if the modeling error ε is used for adaptation, potential improvements can be obtained. We now introduced two methods of alternate adaptive control in lieu of the existing method.

3.1 Direct Adaptive Control with Recursive Least Squares

In this approach, we will design a neural net weight update law that minimizes the non-homogeneous term in Eq. (6). We will use the optimal estimation method to minimize the following cost functional

$$J\left(\mathbf{W}\right) = \frac{1}{2}\int_0^t \left(1 + \boldsymbol{\beta}^\top \mathbf{R}\boldsymbol{\beta}\right)^{-1} \left\|\mathbf{W}^\top \boldsymbol{\beta} - \hat{\varepsilon}\right\|^2 d\tau \tag{52}$$

where

$$\hat{\varepsilon} = \varepsilon - \Delta\varepsilon = \dot{\hat{\omega}} - \dot{\omega}_d = \dot{\hat{\omega}} - \mathbf{F}_1^*\omega - \mathbf{F}_2^*\sigma - \mathbf{G}^*\hat{\delta} \tag{53}$$

$$\xi = \boldsymbol{\beta}^\top \mathbf{R}\boldsymbol{\beta} \tag{54}$$

$\Delta\varepsilon$ is the estimation error of the modeling error ε since aircraft angular accelerations $\dot{\omega}$ may not be directly measured and thus requires to be estimated. ξ is viewed as a weighted PE condition that is required for improved convergence with $\mathbf{R}$ as a positive-definite weighting matrix. A larger $\mathbf{R}$ results in a faster convergence. $\mathbf{R}$ can also be viewed as a learning rate matrix.

To minimize the cost functional, we compute its gradient with respect to the neural net weight matrix and set it to zero, thus resulting in

$$\nabla J_{\mathbf{W}}^\top = \int_0^t (1+\xi)^{-1} \boldsymbol{\beta}\left(\boldsymbol{\beta}^\top \mathbf{W} - \hat{\varepsilon}^\top\right) d\tau = \mathbf{0} \tag{55}$$

Equation (55) is then written as

$$\int_0^t (1+\xi)^{-1} \boldsymbol{\beta}\boldsymbol{\beta}^\top d\tau \mathbf{W} = \int_0^t (1+\xi)^{-1} \boldsymbol{\beta}\hat{\varepsilon}^\top d\tau \tag{56}$$

Let

$$\mathbf{R}^{-1} = \int_0^t (1+\xi)^{-1} \boldsymbol{\beta}\boldsymbol{\beta}^\top d\tau \tag{57}$$

We then note that

$$\mathbf{R}^{-1}\mathbf{R} = \mathbf{I} \Rightarrow \dot{\mathbf{R}}^{-1}\mathbf{R} + \mathbf{R}^{-1}\dot{\mathbf{R}} = \mathbf{0} \tag{58}$$

Solving for $\dot{\mathbf{R}}$ yields

$$\dot{\mathbf{R}} = -\mathbf{R}\dot{\mathbf{R}}^{-1}\mathbf{R} = -\left(1+\xi\right)^{-1}\mathbf{R}\boldsymbol{\beta}\boldsymbol{\beta}^{\top}\mathbf{R} \tag{59}$$

Differentiating Eq. (56) yields

$$\mathbf{R}^{-1}\dot{\mathbf{W}} + \left(1+\xi\right)^{-1}\boldsymbol{\beta}\boldsymbol{\beta}^{\top}\mathbf{W} = \left(1+\xi\right)^{-1}\boldsymbol{\beta}\hat{\varepsilon}^{\top} \tag{60}$$

Solving for $\dot{\mathbf{W}}$ yields the neural net weight update law

$$\dot{\mathbf{W}} = -\left(1+\xi\right)^{-1}\mathbf{R}\boldsymbol{\beta}\left(\boldsymbol{\beta}^{\top}\mathbf{W} - \hat{\varepsilon}^{\top}\right) \tag{61}$$

Equation (61) is a gradient-based recursive least-squares neural net weight update law that minimizes the neural net approximation error. In the process, the tracking error should decrease optimally to some minimum convergence radius. Comparing this neural net weight update law to that in Eq. (9), it is seen that the estimated modeling error $\hat{\varepsilon}$ is used for adaptation instead of tracking error. Moreover, the learning rate is also adaptive in that the weighting matrix $\mathbf{R}$ also needs to be updated by Eq. (59).

The recursive least-squares neural net weight update law can be shown to be stable and result in bounded signals. To show this, let $\mathbf{W} = \mathbf{W}^{*} + \tilde{\mathbf{W}}$ with the asterisk and tilde symbols denoting ideal neural net weight matrix and neural net weight matrix deviations, respectively. The following Lyapunov candidate function is chosen

$$\dot{V} = \mathbf{e}^{\top}\mathbf{P}\mathbf{e} + \mathrm{tr}\left(\tilde{\mathbf{W}}^{\top}\mathbf{R}^{-1}\tilde{\mathbf{W}}\right) \tag{62}$$

The time rate of change of the Lyapunov candidate function is computed as

$$\dot{V} = \dot{\mathbf{e}}^{\top}\mathbf{P}\mathbf{e} + \mathbf{e}^{\top}\mathbf{P}\dot{\mathbf{e}} + \mathrm{tr}\left(2\tilde{\mathbf{W}}^{\top}\mathbf{R}^{-1}\dot{\tilde{\mathbf{W}}} + \tilde{\mathbf{W}}^{\top}\dot{\mathbf{R}}^{-1}\tilde{\mathbf{W}}\right) \tag{63}$$

Upon simplification, one obtains

$$\dot{V} = -\mathbf{e}^{\top}\mathbf{Q}\mathbf{e} + 2\mathbf{e}^{\top}\mathbf{P}\mathbf{B}\left(\boldsymbol{\psi} + \boldsymbol{\Delta}_1\right) - \left(1+\xi\right)^{-1}\boldsymbol{\psi}^{\top}\boldsymbol{\psi} + 2\left(1+\xi\right)^{-1}\boldsymbol{\Delta}_1^{\top}\boldsymbol{\psi} \tag{64}$$

Then, it can be shown that the tracking error is uniformly ultimately bounded if the following condition is satisfied

$$\lambda_{min}\left(\mathbf{Q}\right)\left\|\mathbf{e}\right\|^{2} - 2\left\|\mathbf{e}\right\|\lambda_{max}\left(\mathbf{P}\right)\left\|\mathbf{B}\right\|\left(\left\|\boldsymbol{\psi}\right\| + \left\|\boldsymbol{\Delta}_1\right\|\right) + \left(1+\xi\right)^{-1}\left\|\boldsymbol{\psi}\right\|\left(\left\|\boldsymbol{\psi}\right\| - 2\left\|\boldsymbol{\Delta}_1\right\|\right) \geq 0 \tag{65}$$

Let $\mathbf{R} = \Gamma \mathbf{I}$, then the system dynamics of the recursive least-squares method are bounded by the following equation

$$\begin{bmatrix} \dot{\mathbf{e}} \\ \dot{\boldsymbol{\psi}} \\ \dot{\mathbf{z}} \end{bmatrix} \leq \begin{bmatrix} \mathbf{A} & \mathbf{B} & \mathbf{B} \\ \mathbf{0} & -\left(1+\xi\right)^{-1} \mathbf{R}\alpha & \mathbf{0} \\ \frac{\eta}{\epsilon}\mathbf{G}\mathbf{G}^{*-1}\mathbf{B}^{\top}\mathbf{A} & \frac{\eta}{\epsilon}\mathbf{G}\mathbf{G}^{*-1} & -\frac{1}{\epsilon} \end{bmatrix} \begin{bmatrix} \mathbf{e} \\ \boldsymbol{\psi} \\ \mathbf{z} \end{bmatrix} + \begin{bmatrix} \boldsymbol{\Delta}_1 \\ \boldsymbol{\Delta}_2 \\ \boldsymbol{\Delta}_3 \end{bmatrix} \tag{66}$$

Let $\mathbf{A} = -\mathbf{K}_p$, $\mathbf{G} = \mathbf{G}^*$, $\mathbf{B} = \mathbf{I}$, so that $\mathbf{B}^{\top}\mathbf{P} = \mathbf{P}_{12}$. Then, the characteristic equation is evaluated as

$$\det \begin{bmatrix} s + \mathbf{K}_p & -\mathbf{I} & -\mathbf{I} \\ 0 & s\mathbf{I} + \left(1+\xi\right)^{-1} \mathbf{R}\alpha & \mathbf{0} \\ \frac{\eta}{\epsilon}\mathbf{K}_p & -\frac{\eta}{\epsilon} & \left(s+\frac{1}{\epsilon}\right)\mathbf{I} \end{bmatrix} = \prod_{j=1}^{3} \left[s + \left(1+\xi\right)^{-1} \Gamma\alpha\right] \times \left[\left(s+k_{p,j}\right)\left(s+\frac{1}{\epsilon}\right) + \frac{\eta}{\epsilon}k_{p,j}\right] = 0 \tag{67}$$

Without the effects of unmodeled dynamics, this characteristic equation yields two set of roots, both of which lie on the open left-half s-plane

$$s_{1,j} = -k_{p,j} \tag{68}$$

$$s_{2,j} = -\left(1+\xi\right)^{-1} \Gamma\alpha \approx -\frac{\Gamma\alpha}{1+\Gamma\alpha} \tag{69}$$

Since $s_{2,j}$ is a negative real root corresponding to the neural net weight matrix variation, increasing the learning rate would simply drive $s_{2,j}$ farther to the left on the real axis of the s-plane. Therefore, unlike the current neural net direct adaptive control approach, large learning rate does not generate high frequency oscillations with the recursive least-squares method. Also, the rate of convergence tends to unity for large learning rate.

With the effects of unmodeled dynamics, one set of roots of the characteristic equation is given by Eq. (69) and the other set of roots are

$$s_{1,2,j} = -\frac{k_{p,j}+\frac{1}{\epsilon}}{2} \pm \frac{1}{2}\sqrt{\left(k_{p,j}-\frac{1}{\epsilon}\right)^2 - 4\frac{k_{p,j}\eta}{\epsilon}} \approx -\frac{k_{p,j}+\frac{1}{\epsilon}}{2} \pm \frac{-k_{p,j}+\frac{1}{\epsilon}}{2} \approx -k_{p,j},\ -\frac{1}{\epsilon} \tag{70}$$

All the roots are thus stable without any condition on the learning rate. Therefore, the recursive least-squares method would tend to be less sensitive to unmodeled dynamics than the current neural net direct adaptive control.

As can be seen from Fig. 8 , the recursive least-squares learning provides a significant improvement in the tracking performance of the adaptive control. Moreover, increasing the learning rate does not cause high frequency oscillations as in the case of the current direct adaptive control approach. This is

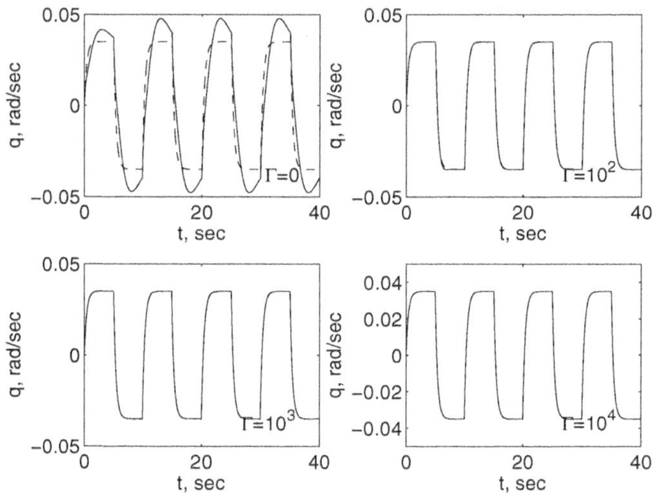

Fig. 9 Pitch Rate with RLS Method

in agreement with the analysis which shows that the root s_2 corresponding to the neural net weight update law does not have an imaginary part.

Fig. 10 is the plot of the selected neural net weights. The weights exhibit a nice convergence behavior. Increasing the learning rate causes the weight to move closer to the true values of the system parameters for which the adaptive control is compensating. In contrast, the neural net weights in the current adaptive control approach do not converge correctly to their true values, as shown in Fig. 5 .

Fig. 11 shows the recursive least-squares learning in the presence of unmodeled dynamics. In contrast with the current method as shown in Fig. 8, the recursive least-squares learning is able to handle unmodeled dynamics much better. Increasing the learning rate does not cause increased high frequency oscillations. So the sensitivity to unmodeled dynamics is much less of an issue with the recursive least-squares learning. This behavior is in agreement with the analysis.

3.2 Hybrid Direct-Indirect Adaptive Control with Recursive Least-Squares

Another adaptive control architecture that has recently been proposed is hybrid adaptive control (20). This architecture is as shown in Fig. 12. The hybrid adaptive control blends both direct and indirect adaptive control methods together to provide a more effective control strategy. The indirect adaptive control is responsible for updating the dynamic inversion controller

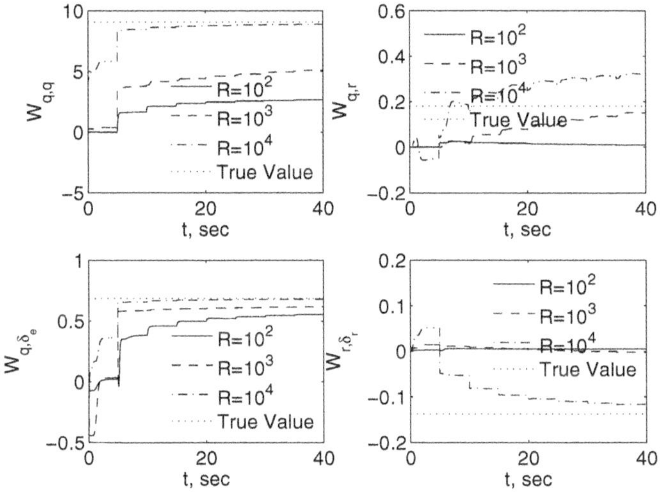

Fig. 10 Neural Net Weight Learning with RLS Method

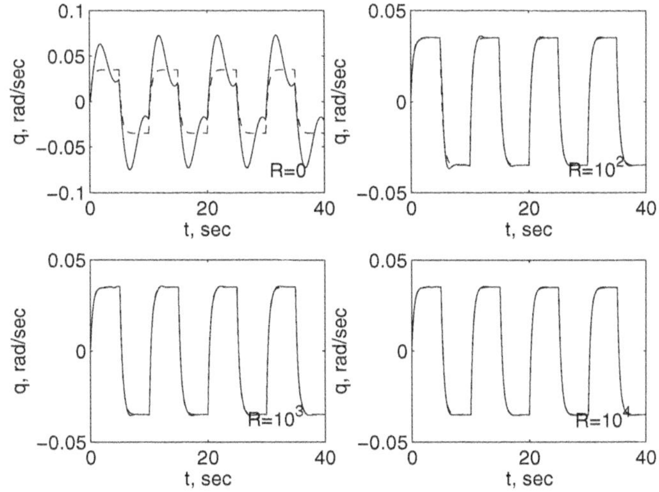

Fig. 11 Pitch Rate with RLS Method and Unmodeled Dynamics ($\epsilon = 0.1$, $\eta = 0.1$)

with a more accurate plant model which is estimated by the recursive least squares method. Any residual tracking error as a result of the dynamic inversion can then be handled by the neural net direct adaptive control.

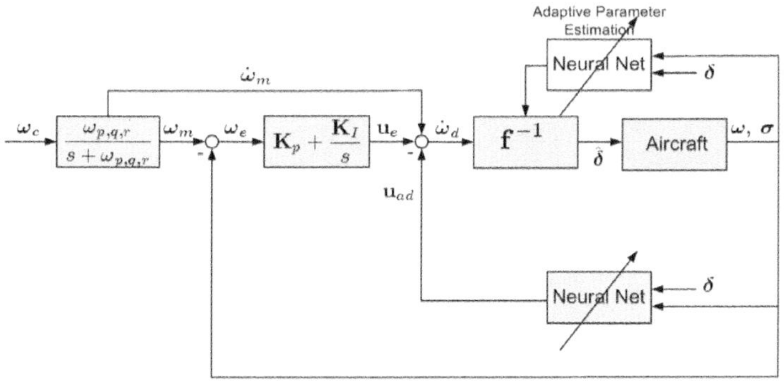

Fig. 12 Hybrid Adaptive Flight Control

The dynamic inversion controller is updated by the estimated plant model at every time step according to

$$\delta = \hat{\mathbf{G}}^{-1}\left(\dot{\omega}_d - \hat{\mathbf{F}}_1\omega - \hat{\mathbf{F}}_2\sigma\right) \tag{71}$$

where $\hat{\mathbf{F}}_1 = \mathbf{F}_1^* + \Delta\hat{\mathbf{F}}_1$, $\hat{\mathbf{F}}_2 = \mathbf{F}_2^* + \Delta\hat{\mathbf{F}}_2$, $\hat{\mathbf{G}} = \mathbf{G}^* + \Delta\hat{\mathbf{G}}$ are the estimated plant matrices of the true plant model and $\hat{\mathbf{G}}$ is assumed to be invertible.

The hybrid adaptive control performs explicit parameter identification of the plant model to account for changes in aircraft dynamics. The parameter identification process is performed by the recursive least-squares method

$$\dot{\boldsymbol{\Phi}} = -\left(1+\xi\right)^{-1}\mathbf{R}\boldsymbol{\theta}\left(\boldsymbol{\theta}^\top\boldsymbol{\Phi} - \hat{\varepsilon}^\top\right) \tag{72}$$

where $\boldsymbol{\Phi}^\top = \left[\mathbf{W}_\omega^\top\ \mathbf{W}_\sigma^\top\ \mathbf{W}_\delta^\top\right]$ is a neural net weight matrix and $\boldsymbol{\theta}^\top = \left[\boldsymbol{\omega}^\top\boldsymbol{\beta}_\omega^\top\ \boldsymbol{\sigma}^\top\boldsymbol{\beta}_\sigma^\top\ \boldsymbol{\delta}^\top\boldsymbol{\beta}_\delta^\top\right]$ is an input matrix.

The estimated plant matrices are then updated as

$$\hat{\mathbf{F}}_1 = \mathbf{F}_1^* + \mathbf{W}_\omega^T\boldsymbol{\beta}_\omega \tag{73}$$

$$\hat{\mathbf{F}}_2 = \mathbf{F}_2^* + \mathbf{W}_\sigma^T\boldsymbol{\beta}_\sigma \tag{74}$$

$$\hat{\mathbf{G}} = \mathbf{G}^* + \mathbf{W}_\delta^T\boldsymbol{\beta}_\delta \tag{75}$$

The performance of the hybrid adaptive control is very similar to the recursive least-squares direct adaptive control with large values of $\mathbf{R}$. At smaller values of $\mathbf{R}$, the adaption is shared between the neural net direct and indirect adaptive control blocks. Thus, the learning rate of the neural net direct adaptive control can be turned down to reduce potential excitation of unmodeled dynamics as discussed earlier. The advantage of the hybrid adaptive control method is the ability to be able to estimate plant model parameters on-line. Direct adaptive control approaches accommodate changes in plant dynamics implicitly but do not provide an explicit means for ascertaining the knowledge of the plant dynamics. By estimating the plant model parameters explicitly using the recursive least-squares neural net learning law, an improved knowledge of the plant dynamics can be obtained that can potentially be used to develop fault detection and isolation (FDI) strategies, and emergency flight planning to provide guidance laws for energy management in the presence of hazards.

4 Verification and Validation Challenges for Adaptive Systems

Creating certifiable adaptive flight control systems represents a major challenge to overcome. Adaptive control systems with learning algorithms will never become part of the future unless it can be proven that this software is highly safe and reliable. Rigorous methods for adaptive software verification and validation must therefore be developed by NASA and others to ensure that control system software failures will not occur, to verify that the control system functions as required, to eliminate unintended functionality, and to demonstrate that FAA certification requirements can be satisfied.

The ability of an adaptive control system to modify a pre-designed flight control system is at the same time a strength and a weakness. On the one hand, the premise of being able to accommodate vehicle degradation is a major selling point of adaptive control since traditional gain-scheduled control methods are viewed to be less capable of handling off-nominal flight conditions outside their design operating points. Nonetheless, gain-scheduled control approaches are robust to disturbances and secondary dynamics. On the other hand, as previously shown in this paper, potential problems with adaptive control exist with regards to high-gain learning and unmodeled dynamics. Clearly, adaptive control algorithms are sensitive to these potential problems as well as others that have not been considered such as actuator dynamics, exogenous disturbances, etc. Moreover, a certifiable adaptive flight control law must be able to accommodate these effects as well as other factors such as time delay, system constraints, and measurement noise in a globally satisfactory manner.

4.1 Simulation of Adaptive Control Systems

Simulation will likely continue to play a major role in the verification of learning systems. Although many advanced techniques, such as model checking, have been developed for finite state systems, there application to hybrid adaptive systems in very limited (26; 27). Many aspects of adaptive systems learning, in particular convergence and stability, can only be analyzed with simulation runs that provide enough detail and fidelity to model significant nonlinear dynamics. For example, stall upsets of an aircraft cannot be expressed as a linear model since this effect is highly nonlinear and unsteady. Simulation provides a fairly rapid way to accomplish the following tasks:

- Evaluation and comparison of different learning algorithms.
- Tuning control gains and learning of weight update law.
- Determination of how much learning is actually accomplished at each step.
- Evaluation of the effect of process and measurement noise on learning convergence rate.
- Determination of learning stability boundaries.
- Testing algorithm execution speed on actual flight computer hardware.
- Conducting piloted evaluation of the learning system in a flight simulator.
- Simulating ad-hoc techniques of improving the learning process, such as adding persistent excitation to improve identification and convergence, or stopping the learning process after error is less than a specified error, or after a specified number of iterations.

Simulations differ primarily in the fidelity with which the plant is modeled. Higher fidelity simulations require more complicated mathematical models of the adaptive system and also a greater use of actual (and expensive) controller hardware. In order to be cost-effective, the lowest fidelity testbed are usually used as much as possible. The behavior of simple linear models are compared to that of higher fidelity nonlinear models when they are available to ensure that analysis performed using the linear model still applies. Table 1 presents one representation of the simulation hierarchy from lowest to highest fidelity.

The lowest fidelity simulations are usually run on a desktop computer in the Matlab/Simulink environment. This simulation typically includes the control laws and a linear plant which accounts for the aircraft aerodynamics, mass properties, and engine thrust model. The linear model is most often used in early control law design and analysis or to calculate linear gain and phase margins. It is important to note that nonlinear adaptive controllers can be represented linearly using the error bounded analysis as shown above, but the linear model may not provide results with the required accuracy. Nonetheless, the linear model can provide a very useful insight to the stability and convergence of the nonlinear adaptive controllers. Changes to the plant model can be simulated by changing the system transfer function from one matrix to another with varying frequency. By varying the amount of change, the stability boundaries of the system can be determined. Concomitant with this process is an evaluation of the system tuning parameters that are used in the learning algorithm. The desktop simulation environment provides a quick way to compare different learning algorithms and controller architectures. Only the most promising designs need be simulated using higher fidelity simulations.

Higher fidelity simulation testbeds use actual flight hardware (or even aircraft) in the simulation of the control loop, and are often run in dedicated

Table 1 Simulation Environments

Testbed	Pilot Interface Fidelity	Model Fidelity	Test Environment
Desktop Computer	Low	Low	Linear/nonlinear models using Matlab or Simulink
Workstation	Low	Low to Medium	Can interface with high-fidelity modeling tools
Fixed-Based Simulator	Low to Medium	Medium	Dedicated aircraft model and hardware
Hardware-in-the Loop Simulator	Medium to High	Medium to High	Actual aircraft target flight computer and cockpit
Aircraft-In-the-Loop Simulator	High	Medium to High	Simulator with actual flight computer and ground-based aircraft
Motion-Based Simulator	High	High	Nonlinear simulation with moving cockpit

computing environments with a cockpit and out-the-window graphics (e.g., see (28; 29)). These simulations may include a cockpit to interface with the pilot and can either be fixed-based or motion-based. Motion-based simulators additionally provide the pilot with some of the physical cues of actual flight. Typically they contain software models of nonlinear aerodynamics, engine dynamics, actuator models, and sensor models. The most common elements of these testbeds are some of the flight processors, communication buses and a cockpit. Using the actual aircraft flight computer is a particularly important advantage of this simulation, since all computers tend to handle exceptions differently and may have differences in their numerical routines. Either the actual aircraft may be tied into the nonlinear simulation, or an iron-bird aircraft may be used to provide actuators, sensor noise, actual flight wiring, and some structural interactions. These testbeds allow for a complete check out of all interfaces to the flight hardware, timing tests, and various failure modes and effects analysis (FMEA) testing, which is not possible in a simpler configuration.

4.2 Approach for Adaptive System V&V

The current approach is to verify a neural net adaptive flight control over an exhaustive state space using the Monte Carlo simulation method. The state space must be carefully designed to include all possible effects that an aircraft can encounter in flight. By sensitivity analysis, some of these effects may be considered less significant that could be eliminate to reduce the dimensionality of the state space. For example, aeroelastic effects can be significant for flight vehicles. However, high frequency flexible modes of aircraft structures are generally not easily excitable, thus their effects could be discounted. Other modes, however, may be significant such as those that appear inside the flight control bandwidth. In addition, other dynamical effects should be considered including actuator dynamics, turbulence, sensor noise, digital signal processes that give rise to time delay, etc.

Initial simulations are usually conducted on a desktop PC in the Matlab/Simulink environment. The objective is to test the learning behavior using an ideal model, which may be simply the one used by the theoretical development. Initially, the controller should be operated without any failure nor any learning to acquire baseline performance and to demonstrate controller stability. Once this has been shown, the ability of the control system to learn should be explored. This may be investigated in a two-step process:

1. Initially, a "failure" or step-change is introduced into the system in order to test the learning under ideal conditions. The change could be a change in the A matrix (damage simulation), or a change in the B matrix (actuator failure), or both. In the initial stage, no measurement noise (sensor noise) or process noise (unmodeled dynamics) is introduced. In addition, the controller is allowed to give persistent excitation commands in order to provide the ideal environment for rapid learning convergence. Hence the objective is not to demonstrate controller robustness, but rather only to document how well the learning algorithm can learn under ideal conditions. If the simulation indicates that learning is not occurring even under ideal conditions, then effort should be made to improve the learning by modifying the control law or the neural network architecture.
2. Once the learning under ideal conditions has been judged to be acceptable, the next phase of the learning simulation is to test learning under non-ideal conditions. Measurement noise should be added to the simulation to estimate the level of persistent excitation required to maintain convergent learning. As indicated above by the theory, the learning rate will likely be a function of the level of persistent excitation. In some severe cases, however, the addition of measurement noise may destabilize the learning process even for large amounts of persistent excitation.

Once the learning algorithm is felt to operate successfully in a simulated environment, then the performance of the learning system can be evaluate while using the controller to reject disturbances. These simulations may reveal the necessity to disable learning as the adaptation errors become low. This could be done to prevent the learning algorithm from seeking to map measurement noise to small changes in the control input. The choice of learning rate and neural net weight limits will also be evaluated in simulation to guide gain selection for actual testing. Although higher learning gains tend to increase the speed of learning, high gains also tend to promote instability of the learning algorithm as discussed earlier. Another problem is that defining the stability boundaries of multiple-input, multiple-output adaptive control systems can require many test points at each of many possible operating conditions. For this reason, analytical methods that can determine learning system stability are needed. The analysis presented in this paper can provide an analytical method to help guide the analysis of stability boundaries.

A problem encountered in performing simulation is proving adequate test coverage. Coverage concerns with program execution of flight control software to ensure that its functionality is properly designed. In order to help simulation achieve greater coverage, various tools and methods are being developed to implement simulation in a more systematic manner. One such tool is Automated Neural Flight Controller Test (ANCT) (30) which is developed in the MATLAB environment. ANCT has been designed to help test engineers evaluate different flight conditions, quantify performance, and determine regions of stability. ANCT is designed to analyze a MATLAB/Simulink model using all possible combinations of the model inputs parameters. By introducing random numbers into the test inputs and parameters, a Monte Carlo simulation can be performed to estimate the sets of model parameters and inputs that correspond to the control system responses that are of interest. ANCT evaluates the time-series outputs during a specified time or condition window, and then computes a performance score that represents the degree to which the control system responses meet performance specifications.

Another simulation tool is Robustness Analysis for Control Law Evaluation (RASCLE) which has also been developed to help explore different combinations of learning system parameters and operating conditions (31). RASCLE can interface with existing nonlinear simulations and incorporates search algorithms to uncover regions of instability with as few runs as possible. RASCLE uses a gradient algorithm to identify the direction in the uncertainty space along which the stability of the control system is most rapidly decreasing. RASCLE provides an intelligent simulation-based search capability that can be used in Monte Carlo simulation evaluations (32).

5 Future Research

5.1 Adaptive Control

Despite the extensive progress made in adaptive control research from the 1970's until the present time, this technology has not been adopted for use in primary flight control systems in mission-critical or human-rated flight vehicles.The following quote from the IRAC Project Proposal (3) highlights the challenges with adaptive control: " In 2004 a NASA Aeronautics "Adaptive Controls Task Force" with representation from NASA Ames, Dryden, Glenn, and Langley observed that existing flight control technology is not adequate to handle large uncertainties and system changes, unknown component failures and anomalies, high degree of complexity, non-linear unsteady dynamics, revolutionary vehicles, and novel actuators and sensors. The Task Force further observed that uncertainties and system changes can be continuous or discrete, such as varying flight conditions, abrupt failures, and structural damage, to name a few."

The existing approach to adaptive control synthesis generally lacks the ability to deal with integrated effects of many different flight physics as pointed out above. In the presence of hazards such as damage or failures, flight vehicles can exhibit numerous coupled effects such as aerodynamics, vehicle dynamics, structures, and propulsion. These coupled effects impose a considerable amount of uncertainties on the performance of a flight control system. Thus, even though an adaptive control may be stable in a nominal flight condition, it may fail to maintain enough control margins in the presence of these uncertainties. For example, conventional aircraft flight control systems incorporate aeroservoelastic filters to prevent control signals from exciting wing flexible modes. If changes in the aircraft configuration are significant enough, frequencies of the flexible modes may be shifted that render the filters ineffective. This would allow control signals to potentially excite flexible modes which can cause problems for a pilot to maintain good tracking control. Another example is the use of slow actuators such as engines as control effectors. In off-nominal events, engines are sometimes used to control aircraft. This has been shown to enable pilots to maintain control in some emergency situations such as the DHL incident involving an Airbus A300-B4 in 2003 that suffered structural damage and hydraulic loss over Baghdad (33), and the Sioux City, Iowa accident involving United Airlines Flight 232 (1). The dissimilar actuator rates can cause problems with adaptive control and can potentially lead to pilot-induced oscillations (PIO) (34).

To adequately deal with these coupled effects, an integrated approach in adaptive control research should be taken. This integrated approach will require developing new fundamental multidisciplinary methods in adaptive control and modeling. As discussed earlier, unmodeled dynamics are a source of significant uncertainties that can cause an adaptive control algorithm to become unstable if high-gain learning is used. Thus, a multidisciplinary approach in adaptive control research would be to develop fundamental understanding of the structures of these secondary dynamics which would bring together different disciplines such as aerodynamics and structures. With a better understanding of the system uncertainties, more effective adaptive control methods could be developed to improve robustness in the presence of uncertainties.

Another future research goal is to extend the concept of linear control margins to adaptive control disciplines. Adaptive control methods are generally time-domain methods since Lyapunov analysis works in time domain. Yet, robust control is usually done in the frequency domain. Robust control requires a controller to be analyzed using the phase and gain margin concepts in the frequency domain. With this tool, an adaptive control can be analyzed to assess its control margin sensitivity for different learning rates. This would then enable a suitable learning rate to be determined. By incorporating the knowledge of unmodeled dynamics, a control margin can be evaluated to see if it is sufficient to maintain stability of a flight control system in the presence of potential hazards.

5.2 Verification and Validation

Verification and validation research is viewed as a key research to enable adaptive control to be operational in future flight vehicles. V&V processes are designed to ensure that adaptive systems function as intended and the consequences of all possible outcomes of the adaptive control are verified to be acceptable. Software certification is a major issue that V&V research is currently addressing. Some of the future research in software certification for adaptive control are discussed as follows:

- Model Checking for Hybrid Adaptive System:
  Over the last decade, the formal method of model checking has become an important tool for the verification of finite state automata. Model checkers have found considerable applications for outer-loop adaptive control system verification. They have been useful for verification of autonomous systems such as NASA Remote Agent and K9 Mars Rover (35), and by Rockwell Collins to provide verification of the mode logic of the FCS 5000 flight guidance system being developed for use in business and regional jet aircraft (36). The outer-loop controller of these programs use planners and schedulers to coordinate the actions of multiple program threads that execute in parallel. A future challenge is to extend the technique of model checking to verification of inner-loop control and learning adaptation. These processes are generally continuous systems, not finite state automata. Nevertheless, some recent progress has been made attempting to apply the technique of hybrid model checking to continuous systems. Ref (27) describes an application of Java PathFinder to the control of a robotic vehicle. The vehicle dynamics are modeled in the time domain as a set of first order differential equations. The execution of the inner-loop controller is controlled by an outer-loop autonomous agent planner and scheduler. Although the continuous variables could assume an infinite number of values, thereby presenting a state explosion problem for the model checker, the use of Java PathFinder is made possible through representing theses values as discrete quantities. The use of an approximation function converts the continuous variables into discrete values. The idea is similar to rounding a decimal number to the nearest integer, only in this case, the truncation is considerably coarser. With this abstraction of the continuous space, the variables can be made to take on relatively few values. This allows for the recognition of previous "states" in the model checking sense of the word, and hence an exploration of the continuous model checking space becomes possible. Of course, this search is exhaustive only to the

extent the approximation function is valid. If the approximation function is too coarse, important states will likely be missed.

- Program Synthesis Methods for Certifiable Code Generation:
  In the future, it may be possible to use software tools to help produce certifiable code, including code for learning systems. Although software produced by these tools would still undergo a formal certification process, the idea is to generate certificates automatically together with the software. As an example, AutoFilter is a tool being developed at NASA Ames to automatically generate certifiable Kalman Filter code from high-level declarative specifications of state estimation problems (37). Although Kalman filters are widely used for state estimation in safety-critical systems, the complex mathematics and choice of many tuning parameters make implementation a difficult task. The AutoFilter tool not only generates Kalman filter code automatically from high level specifications, but also generates various human-readable documents containing both design and safety related information required by certification standards. Program synthesis is accomplished through repeated application of schemas, or parametrized code fragment templates and a set of constraints formalizing the template's applicability to a given task. Schemas represent the different types of learning algorithms. AutoFilter applies rules of the logic backwards and computes, statement by statement, logical s or safety obligations which are then processed further by an automatic theorem prover. To perform this step automatically, however, auxiliary annotations are required throughout the code. AutoFilter thus simultaneously synthesizes the code and all required annotations. The annotations thereby allow automatic verification and produces machine-readable certificates showing that the generated code does not violate the required safety properties.
- Tools for On-line Software Assurance:
  Although simulation test cases may discover problems, testing can never reveal the absence of all problems, no matter how many high-fidelity simulations are performed. It is for this reason that undiscovered failure modes may lurk in the control system or be found at a test condition previously not simulated. To safeguard against these failures, means of verifying in-flight software assurance should be developed. As one approach to this problem, NASA Ames has developed a tool called the Confidence Tool to analyze the probability distribution of the neural network output using a Bayesian approach (38). This approach combines mathematical analysis with dynamic monitoring to compute the probability density function of neural network outputs while the learning process is on-going. The Confidence Tool produces a real-time estimate of the variance of the neural network outputs. A small variance indicates the network is likely producing a good, reliable estimate, and therefore, good performance of the neural network software can be expected. The confidence tool can be used for pre-deployment verification as well as a software harness to monitor quality of the neural network during flight. The outputs of the Confidence

Tool might be used as a signal to stop and start neural network adaptation or be used to provide a guarantee of the maximum network error for certification purposes.

6 Conclusions

This paper has presented a stability and convergence analysis of a neural net adaptive flight control. An error bound analysis has been introduced that enables a linear dynamics to be extracted from the nonlinear adaptive control algorithm for stability and convergence analysis of the neural net weight update law. The effect of the learning rate has been studied by analysis and confirmed by simulations. It has been shown that high-gain learning will likely result in high frequency oscillations that can excite unmodeled dynamics. For certain classes of unmodeled dynamics, it is possible that a high-gain learning can become unstable. A potential improvement has been presented. This improvement, the recursive least-squares learning law, is based on optimal estimation and uses modeling error for adaptation. The analysis shows that high frequency oscillations can be avoided with this learning law. Furthermore, the effect of unmodeled dynamics has been shown to be less sensitive with this learning law. This paper also has presented some thoughts on the verification and validation approach as an enabling technology that will enable adaptive flight control to be realized in future missions. Current challenges in adaptive control and verification and validation remain to be obstacles to realizing the goal of certifiable adaptive control systems. Future research in adaptive control must be multidisciplinary and integrated to better deal with many sources of uncertainties that arise from coupled effects manifested in flight in the presence of hazards. In this paradigm, adaptive control methods would need to be cognizant of system constraints imposed by dissimilar physical effects while maintaining robustness in the presence of uncertainties. The future research in these disciplines can bear fruits and perhaps will enable adaptive control to be operational someday in the near future.

References

1. National Transportation Safety Board, Aircraft Accident Report. United Airlines Flight 232, McDonnell Douglas DC-10-10, Sioux Gateway Airport, Sioux City, Iowa July 19, 1989, NTSB/AAR-90/06 (1990)
2. National Transportation Safety Board. In-Flight Separation of Vertical Stabilizer American Airlines Flight 587 Airbus Industrie A300-605R, N14053 Belle Harbor, New York, November 12, 2001, NTSB/AAR-04/04 (2004)
3. Totah, J., Krishnakumar, K., Vikien, S.: Integrated Resilient Aircraft Control - Stability, Maneuverability, and Safe Landing in the Presence of Adverse Conditions. NASA Aeronautics Research Mission Directorate Aviation Safety Program, April 13 (2007)

4. Nguyen, N., Krishnakumar, K., Kaneshige, J., Nespeca, P.: Dynamics and Adaptive Control for Stability Recovery of Damaged Asymmetric Aircraft. In: Proc. of AIAA Guidance, Navigation, and Control Conference, AIAA-2006-6049 (2006)
5. Jacklin, S.A., Schumann, J.M., Gupta, P.P., Richard, R., Guenther, K., Soares, F.: Development of Advanced Verification and Validation Procedures and Tools for the Certification of Learning Systems in Aerospace Applications. In: Proc. of Infotech@aerospace Conference, Arlington, VA, September 26-29 (2005)
6. Rohrs, C.E., Valavani, L., Athans, M., Stein, G.: Robustness of Continuous-Time Adaptive Control Algorithms in the Presence of Unmodeled Dynamics. IEEE Transactions on Automatic Control AC-30(9), 881–889 (1985)
7. Steinberg, M.L.: A Comparison of Intelligent, Adaptive, and Nonlinear Flight Control Laws. In: AIAA Guidance, Navigation, and Control Conference, AIAA-1999-4044 (1999)
8. Eberhart, R.L., Ward, D.G.: Indirect Adaptive Flight Control System Interactions. International Journal of Robust and Nonlinear Control 9(14), 1013–1031 (1999)
9. Rysdy, R.T., Calise, A.J.: Fault Tolerant Flight Control via Adaptive Neural Network Augmentation. In: Proc. of AIAA Guidance, Navigation, and Control Conference, AIAA-1998-4483 (1998)
10. Kim, B.S., Calise, A.J.: Nonlinear Flight Control Using Neural Networks. Journal of Guidance, Control, and Dynamics 20(1), 26–33 (1997)
11. Johnson, E.N., Calise, A.J., El-Shirbiny, H.A., Rysdyk, R.T.: Feedback Linearization with Neural Network Augmentation Applied to X-33 Attitude Control. In: Proc. of AIAA Guidance, Navigation, and Control Conference, AIAA-2000-4157 (2000)
12. Narendra, K.S., Annaswamy, A.M.: A New Adaptive Law for Robust Adaptation Without Persistent Excitation. IEEE Transactions on Automatic Control AC-32(2), 134–145 (1987)
13. Krishnakumar, K., Limes, G., Gundy-Burlet, K., Bryant, D.: An Adaptive Critic Approach to Reference Model Adaptation. In: Proc. of AIAA Guidance, Navigation, and Control Conference, AIAA-2003-5790 (2003)
14. Hovakimyan, N., Kim, N., Calise, A.J., Prasad, J.V.R., Corban, E.J.: Adaptive Output Feedback for High-Bandwidth Control of an Unmanned Helicopter. In: Proc. of AIAA Guidance, Navigation and Control Conference, AIAA-2001-4181 (2001)
15. Bishop, C.M.: Neural Networks for Pattern Recognition. Oxford University Press, Oxford (1995)
16. Lewis, F.W., Jagannathan, S., Yesildirak, A.: Neural Network Control of Robot Manipulators and Non-Linear Systems. CRC Press, Boca Raton (1998)
17. Zurad, J.M.: Applications of Neural Networks for Aerospace-Related Technologies. In: Proc. of IEEE Aerospace Applications Conference (1996)
18. Williams-Hayes, P.S.: Flight Test Implementation of a Second Generation Intelligent Flight Control System. Technical Report TR-213669, NASA (2005)
19. Bosworth, J., Williams-Hayes, P.S.: Flight Test Results from the NF-15B IFCS Project with Adaptation to a Simulated Stabilator Failure. In: Proc. of AIAA Infotech@Aerospace Conference, AIAA-2007-2818 (2007)

20. Nguyen, N., Krishnakumar, K.: A Hybrid Flight Control with Adaptive Learning Parameter Estimation. In: Proc. of AIAA Infotech@Aerospace Conference, AIAA-2007-2841 (May 2007)
21. Nguyen, N., Bakhtiari-Nejad, M., Huang, Y.: Hybrid Adaptive Flight Control with Bounded Linear Stability Analysis. In: Proc. of AIAA Guidance, Navigation, and Control Conference, AIAA-2007-6422 (August 2007)
22. Nguyen, N., Boskovic, J.: Bounded linear Stability Margin Analysis of Nonlinear Hybrid Adaptive Control. In: Proc. of the American Control Conference (June 2008)
23. Cao, C., Patel, V.V., Reddy, C.K., Hovakimyan, N., Lavretsky, E., Wise, K.: Are Phase and Time-Delay Margin Always Adversely Affected by High Gains? In: Proc. of AIAA Guidance, Navigation, and Control Conference, AIAA-2006-6347 (August 2006)
24. Ioannu, P.A., Sun, J.: Robust Adaptive Control. Prentice-Hall, Englewood Cliffs (1996)
25. Bailey, R.M., Hostetler, R.W., Barnes, K.N., Belcastro, C.M.: Experimental Validation: Subscale Aircraft Ground Facilities and Integrated Test Capability. In: Proc. of AIAA Guidance, Navigation, and Control Conference, AIAA-2005-6433 (2005)
26. Visser, W., Havelund, K., Brat, G., Park, S., Lerda, F.: Model Checking Programs. Kluwer Academic Publisher, Dordrecht (2002)
27. Scherer, S., Lerda, F., Clarke, E.: Model Checking of Robotic Control Systems. In: Proc. of ISAIRAS Conference, Munich, Germany (2005)
28. Belcastro, C., Belcastro, C.: On the Validation of Safety Critical Aircraft Systems, Part II: Analytical & Simulation Methods. In: Proc. of AIAA Guidance, Navigation, and Control Conference, AIAA-2003-5560 (August 2003)
29. Duke, E.L., Brumbaugh, R.W., Disbrow, D.: A Rapid Prototyping Facility for Flight Research in Advanced Systems Concepts. IEEE Computer (May 1989)
30. Soares, F., Loparo, K.A., Burken, J., Jacklin, S.A., Gupta, P.P.: Verification and Validation of Real-time Adaptive Neural Networks using ANCT Tools and Methodologies. In: Proc. of AIAA Infotech@Aerospace Conference, AIAA-2005-6997 (September 2005)
31. Bird, R.: RASCLE Version 2.0: Design Specification, Programmer's Guide, and User's Guide, Baron Associates, Inc. (February 2002)
32. Belcastro, C., Belcastro, C.: On the Validation of Safety Critical Aircraft Systems, Part I: Analytical & Simulation Methods. In: Proc. of AIAA Guidance, Navigation, and Control Conference, AIAA-2003-5559 (August 2003)
33. Lemaignan, B.: Flying with no Flight Controls: Handling Qualities Analyses of the Baghdad Event. In: Proc. of AIAA Atmospheric Flight Mechanics Conference, AIAA-2005-5907 (August 2005)
34. Gilbreath, G.P.: Prediction of Pilot-Induced Oscillations (PIO) due to Actuator Rate Limiting Using the Open-Loop Onset Point (OLOP) Criterion. M.S. Thesis, Air Force Institute of Technology, Wright-Patterson Air Force Base, Ohio (2001)
35. Giannakopoulou, D., Pasareanu, C., Cobleigh, J.: Assume-Guarantee Verification of Source Code with Design-Level Assumptions. In: Proc. of the 26th International Conference on Software Engineering (ICSE), Edinburgh, Scotland (2004)

36. Miller, S., Anderson, E., Wagner, L., Whalen, M., Heimdahl, M.: Formal Verification of Flight Critical Software. In: Proc. of AIAA Guidance, Navigation, and Control Conference, AIAA-2005-6431 (August 2005)
37. Denney, E., Fischer, B., Schumann, J.M., Richardson, J.: Automatic Certification of Kalman Filter for Reliable Code Generation. IEEE Paper No. 1207: 0-7803-8870 (April 2005)
38. Gupta, P., Schumann, J.M.: A Tool for Verification and Validation of Neural Network Based Adaptive Controllers for High Assurance Systems. In: Proc. of IEEE of High Assurance Software Engineering, HASE (2004)

Dynamic Allocation in Neural Networks for Adaptive Controllers

Sampath Yerramalla, Edgar Fuller, and Bojan Cukic

Abstract. Dynamic allocation in Neural Networks is the process of strategic addition of nodes during the evolution of a feature map. As the trend of using growing neural networks is rising in adaptive controller applications it is important to understand the robustness of the process of dynamic allocation in neural networks. In this paper we analyze the robustness of the process of dynamic allocation that are commonly utilized in growing neural networks to varying, and non-stationary input data. The analysis indicates that dynamic allocation in growing neural networks is not fully robust if based solely on the information from resource values or connectivity structure of the nodes. Based on the observations made, we propose a data-driven dynamic allocation algorithm that is useful for growing neural networks used in adaptive controller applications. The advantage of the proposed algorithm is that it allows neural networks to localize the information represented in the input data while ensuring that the overall topology of the data is preserved. Experimental results are presented to demonstrate using high-dimensional, multivariate data obtained from an adaptive flight controller simulator. The analytical and experimental results affirm the robustness and establish the precedence of the developed dynamic allocation algorithm for adaptive controller applications. We investigate the process of dynamic allocation in the Dynamic Cell Structures neural network algorithm, a representative growing neural network used for on-line learning in adaptive controllers, but the approach presented is applicable to any growing neural network where node insertion is performed in order to improve data modeling.

1 Introduction

Topology preservation is an important aspect of information processing that is applicable in many fields including exploratory data analysis, adaptive control, signal and image processing, and pattern recognition. Topology

J. Schumann, Y. Liu (Eds.): Appl. of Neural Networks in High Assur. Sys., SCI 268, pp. 111–139.
springerlink.com

preservation, in simplified terms can be described as a projection where neighboring or closer features in the input data are mapped onto neighboring output units to the extent possible. Topology preservation ensures that connected structures remain connected and the neighborhood or closeness relationship is retained (12).

Topology preserving neural networks like the Kohonen self-organizing maps (K-SOMs) are characterized by the ability to extract information encoded in the input data of arbitrary high-dimension into features using prototypes, called weight vectors (also, called reference vectors or codebook vectors) (16). Most static neural networks like the K-SOMs require that the network size and shape be predetermined (16). It has been realized that an optimal selection of network size and shape can only be determined after accounting for the statistical properties of the input data (5). In adaptive controller applications, the input data (system parameters being controlled) is non-stationary, and often times there is no prior statistical information about the data. For example, in an adaptive flight control system, the velocity of the aircraft can change drastically with varying frictions and altitudes (4; 14). It then becomes highly desirable in such applications to account for varying, and non-stationary input data by utilizing neural networks that evolve the network size and shape during the formation of the feature map.

Dynamic allocation in neural networks allows the neural network feature map to evolve its structure via strategic addition of nodes and into the specific regions in the neural network. Neural networks that utilize the process of dynamic allocation for evolving their structure are known as Dynamic or Growing neural networks. Growing Cell Structures (GCS), Growing Gas Model, Growing Neural Gas (GNG), Growing Self-Organizing Maps, and Dynamic Cell Structures (DCS) are some of the examples of growing neural networks (7; 10; 11; 13).

In the recent years, there is an increase in the trend of using growing neural networks for on-line learning purposes in safety-critical adaptive controller applications. For example, NASA's first generation Intelligent Flight Control System program utilizes the Dynamic Cell Structures (DCS) neural network for on-line learning and estimation of system parameters (4; 14). The DCS represents a dynamic neural network that combines the concepts of a growing network size with the competitive Hebb learning rule (6; 7; 8; 12). The primary role of a neural network in an adaptive controller is to a generate a feature map of the system parameters (input data) in a topology preserving manner. This means that the neural networks should generate a feature map that closely represents the information in the input data. 'Closeness' can be measured by the amount of topology of the input data that is preserved by the neural network.

In this paper we analyze the robustness of dynamic allocation in neural networks using topology preservation as a heuristic measure of neural network robustness to varying, and non-stationary input data. Using examples of neural network learning from non-stationary input data we show that the commonly utilized approaches of dynamic allocation in neural networks are not fully robust, and can at times generate execution-errors during their operation. Our analysis indicates that dynamic allocation in neural networks should not be completely based on resource values of nodes or its connectivity structure. The analysis indicates that neither the resource value nor the connectivity structure can completely represent the information in the input data at all time.

Based on the observations made during the analysis, we present a dynamic allocation algorithm that is based on addition of nodes directly over input data elements. We provide an analysis to compare the presented approach with the commonly utilized approach. The results from the analysis shows that the proposed approach allows the neural network to continually reduce its approximation error, and thereby reduce the time of the neural network to generate a topology preserving mapping. Using high dimensional, multi-variate data obtained from an experimental adaptive flight control system simulator we demonstrate that the proposed data-driven approach of dynamic allocation is more robust and more reliable in generating topology preserving feature mappings. The analytical and experimental results affirm the applicability of the proposed approach of dynamic allocation for neural networks in adaptive controllers.

1.1 Paper Overview

The paper is organized in the following manner. Sect. 2 provides an analysis of the commonly utilized approach of dynamic allocation in neural networks. In Sect. 3, we describe the steps in the proposed data-driven approach of dynamic allocation. Also in Sect. 3, we provide the analysis to compare the proposed approach of dynamic allocation with the commonly utilized approach. Sect. 4 provides experimental results from the proposed approach of dynamic allocation using high-dimensional, multivariate data obtained from an adaptive flight control system simulator. Finally, conclusions are drawn in Sect. 5 based on the analytical and experimental results to indicate the applicability of the proposed approach of dynamic allocation for neural networks in adaptive controller applications.

2 Dynamic Allocation in Neural Networks

Dynamic allocation allows the size and shape of a neural network to evolve during the generation of a feature map. The process of dynamic allocation in neural networks is commonly based on successive addition of nodes into regions within the neural network. Such neural networks are commonly known as Growing Neural Networks. Current processes of dynamic allocation in neural networks are based on placement of new nodes into regions between the location of nodes with high resource values and their connectivity structure

((5; 7; 10; 13; 15)). For example, in the case of the Dynamic Cell Structures (7) neural networks new nodes are located into regions between the nodes with the highest resource value and its connected node with the highest resource value.

2.1 Dynamic Cell Structures

In 1994, Bruske and Sommer introduced DCS neural network algorithm, a Topology Representing SOM (6; 7; 8). This Topology Preserving Feature Map (TPFM) generation scheme is motivated by Growing Neural GAS algorithm and Topology Representing Networks (TRN) by Martinetz (5; 10; 12; 13; 15).

2.2 Components of DCS Neural Netwrork

As in any SOM-algorithm, DCS neural network is characterized by two essential building blocks: the processing units (neurons) and the lateral connection structure between them. Together they constitute the DCS network. In soft-computing terminology, the processing units are represented by weighted centers ($\mathbf{w}_i \in \mathcal{W}$) and the lateral connections between neural units are represented using connection strengths ($c_{ij} \in \mathcal{C}$).

For instance, consider a DCS network consisting of 3 neurons, numbered as neural units 1, 2, and 3. Let the lateral connection strengths between units 1 and 2, units 2 and 3, units 3 and 1 be $0.7, 0.4, 0.5$, respectively, as shown in Fig. 1.

Then, the weighted center matrix, $\mathcal{W}$ and the connection matrix $\mathcal{C}$, in a 2-dimensional representation, are given as follows.

$$\mathcal{W} = \begin{bmatrix} -1 & 3 \\ 1 & 2 \\ 2 & 3 \end{bmatrix} ; \quad \mathcal{C} = \begin{bmatrix} 0 & 0.7 & 0.5 \\ 0.7 & 0 & 0.4 \\ 0.5 & 0.4 & 0 \end{bmatrix}$$

2.2.1 Network Search

For each randomly drawn input pattern, $\mathbf{m}_i$ from the given training data-set $M \subset \mathbb{R}^D$, DCS algorithm first searches its network for two most receptive (closest) neurons, termed the *Best Matching* and *Second Best* Unit, *BMU* and *SBU*. The algorithm then looks for any existing neighboring neurons connected to the *BMU*. If the *BMU* is not connected to any other neuron of the network, then the *BMU* has no neighbors. The set of all neighboring neurons of the *BMU* is termed as its *neighborhood*, represented by $\{NBR\}$.

$$\{j \in \{NBR\} \quad | \quad c_{ij} \neq 0, \quad i = BMU\}$$

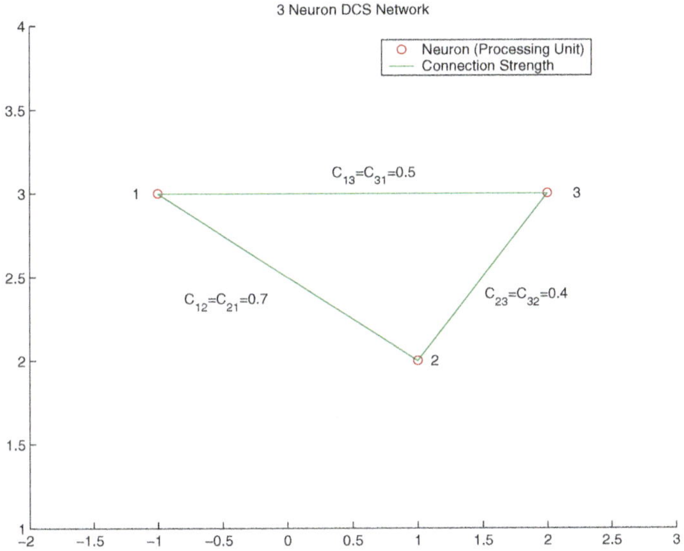

Fig. 1 A 3 Neuron DCS Network

2.2.2 Competitive Hebb Rule (CHR)

DCS neural network rests upon a Radial Basis Function (RBF) and an additional layer of lateral connections between the neural units (6). The lateral connections between neurons are symmetric and bounded in nature, $c_{ij} \equiv c_{ji}, c_{ij} \in [0,1]$. The goal of CHR is to update the lateral connections between the neuron carrying semantic information and its neighboring neurons. By doing so, DCS avoids any restrictions on the topology of the network (7; 8; 12).

After searching the network for the appropriate *BMU*, *SBU*, and the neighborhood $\{NBR\}$ for any randomly drawn input pattern, CHR operates by first setting the connection between the *BMU* and *SBU* to a highest possible strength, $c_{[BMU,SBU]} = 1$. CHR then proceeds by decreasing the strength of any existing connections between members of $\{NBR\}$ and the *BMU* using a forgetting constant, α, as follows

$$c_{[BMU,\{NBR\}]}(t+1) = \alpha c_{[BMU,\{NBR\}]}(t).$$

If any of the existing connections ($c_{[BMU,\{NBR\}]}$), drop below a predefined threshold θ, they are set to zero. Such neurons eventually become non-members of the neighborhood of *BMU*,

$$\textit{if} \quad c_{[BMU,\{NBR\}]} \leq \theta, \quad \textit{then} \quad c_{[BMU,\{NBR\}]} = 0.$$

Thus, CHR can be summarized as follows:

$$c_{ij}(t+1) = \begin{cases} 1 & (i = BMU) \wedge (j = SBU) \\ 0 & (i = BMU) \wedge (j \in \{\{NBR\}\}, j \neq SBU) \\ & \quad \wedge (c_{ij} < \theta) \\ \alpha c_{ij}(t) & (i = BMU) \wedge (j \in \{\{NBR\}\}, j \neq SBU) \\ & \quad \wedge (c_{ij} \geq \theta) \\ c_{ij}(t) & i, j \neq BMU \end{cases}$$

In this manner, CHR induces a Delaunay triangulation into the network by preserving the neighborhood structure of the training data-set and mapping input patterns to relatively close neurons of the network. It was proved in (12) that SOM algorithms utilizing CHR to update lateral connections between neural units carrying semantic information, generate a Topology Preserving Feature Map (TPFM).

2.2.3 Kohonen-Like Neuron Adaptation

As discussed earlier, weights represents the location of the neural units (neurons) of the DCS network, $\mathbf{w}_i \in \mathcal{W} \subset \mathbb{R}^D$. While CHR preserves the *topology*, the weighted centers or the neuron's location in space are updated in a manner that preserves the *geometry* of the training dataset, $M \subset \mathbb{R}^D$. This can be achieved by moving the weighted center of the best matching unit, $\mathbf{w}_{BMU}$ and its neighboring neurons $\mathbf{w}_{NBR}$ closer to the selected input pattern. Thus, for each randomly drawn element, $\mathbf{m}_i \in M$, DCS adapts the weights of the corresponding BMU and its neighborhood set $\{NBR\}$ (6; 7; 8; 9).

Over any training cycle, let $\Delta \mathbf{w}_i = \mathbf{w}_i(t+1) - \mathbf{w}_i(t)$ represent the adaptation of the weight center of a neural unit. Then, the Kohonen-like rule followed in DCS can be summarized as follows:

$$\Delta \mathbf{w}_i = \begin{cases} \varepsilon_{BMU}(\mathbf{m}_i - \mathbf{w}_i(t)) & \textit{if} \quad i = BMU \\ \varepsilon_{NBR}(\mathbf{m}_i - \mathbf{w}_i(t)) & \textit{if} \quad i \in \{NBR\} \\ 0 & \textit{else} \end{cases}$$

where $\varepsilon_{BMU}, \varepsilon_{NBR} \in [0, 1]$ are predefined constants, known as *learning rates* that define the momentum of the learning process. For each input element, $\mathbf{m}_i \in M$, applying CHR before any other adjustment ensures that the SBU is a member of the $\{NBR\}$ set for all further adjustments within the inner loop of the DCS algorithm.

2.2.4 Growing the Network

Unlike traditional SOMs, DCS network has the ability to grow or shrink the map by incrementing or decrementing the number of neurons. A local error measure associated with the network, called *resource* τ, is used to determine if the network experiences a large enough cumulative error. In such a case additional neurons are added into the DCS network. Euclidean distance between the best matching unit (*BMU*) and the presented input element ($\mathbf{m}_i$) is an adequate measure of the resource in most cases,

$$\tau_{BMU} = \|\mathbf{m}_i - \mathbf{w}_{BMU}\|.$$

Decreasing the resource values of all the neurons by a decay constant, β, prevents the resource values from growing out of bounds. After a cycle of adaptation (a learning epoch), the network's *Quantization Error* (QE) is computed as

$$\begin{aligned} \mathrm{QE} &= \sum_{\mathbf{m}_i \in M} \frac{\tau_{BMU}}{N} \\ &= \sum_{\mathbf{m}_i \in M} \frac{\|\mathbf{m}_i - \mathbf{w}_{BMU}\|}{N}, \end{aligned}$$

where N represents the number of neurons in the DCS network at that epoch. Thus, an additional neuron (if needed) is introduced into the network between neurons with highest and second highest resource values, τ_{lg}, τ_{sl}, respectively.

2.3 DCS Algorithm

In DCS algorithm, while the competitive Hebb rule is used to update lateral connections between the neuron carrying semantic information (*BMU, SBU,* $\{NBR\}$), Kohonen-like adaptation is used to shift the weighted centers, $\mathbf{w}_i \in W$, of these neurons closer to the randomly selected input element ($\mathbf{m}_i \in M$). DCS algorithm produces an effective network representation that preserves the features of the input data manifold. Fig. 2 depicts the flow diagram of the DCS neural network algorithm.

The two competitive adaptation rules of the DCS algorithm (CHR and Kohonen-like rule) are the key to its performance during its generation of a TPFM. As shown in Fig. 3, the DCS algorithm is allowed to train on the input data-set until the network has reached a specific stopping criteria.

In summary, after presentation of a random input element ($\mathbf{m}_i \in M \subset \mathbb{R}^D$), the DCS algorithm searches the network for two closest neurons (*BMU,SBU*) and the neighborhood, $\{NBR\}$. The lateral connection structure surrounding the best matching unit, $c_{[BMU,j]}$, is updated using CHR. Kohonen-like adaptation of the weights of the best matching unit ($\mathbf{w}_i | i = BMU$) and its neighbors ($\mathbf{w}_j | j \in \{NBR\}$) is then performed to move these neurons closer to the presented input, $\mathbf{m}_i \in M$. The resource value of the best matching unit, τ_{BMU}, is updated correspondingly, marking the end of

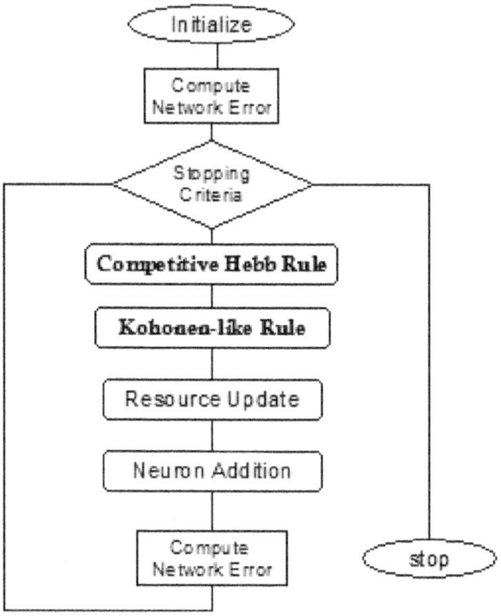

Fig. 2 DCS NN Algorithm Flow Diagram

```
while  stopping criteria is not satisfied
       for each training input stimulus
       {
               find BMU, SBU
               update connections using CHR
               adapt weights using KLR
               update resource error
       }
       compute cumulative network resource error
       if (cumulative network resource error) > (Predefined Error)
       {
               grow the network
               decrement all resource values
       }
}
```

Fig. 3 DCS Algorithm

a training cycle (epoch). The error for the network is computed and used to determine if there is a need for inserting additional neurons into the network.

3 Robustness Analysis of Dynamic Allocation

This type of adaptive behavior raises other questions. Specifically, implementations of software components in high reliability systems must be robust. In other words we require that the software perform consistently when presented with diverse operating conditions, variable architectures or variability in data.

3.1 Node Insertion

During the training process, a neural network is presented each time with a different element of the input data. The feature map is evolved for each input data element by adjusting the weight vectors and the connectivity structure of the neural network based on certain learning algorithms. In the case of the DCS, a Kohonen-like adaptation rule is used for adjusting the weight vectors and a competitive Hebb rule is used for updating the connectivity structure between nodes (7; 11; 13). After a certain number of learning-iterations, which is usually a fixed number in the case of off-line learning, the dynamic allocation algorithm adds nodes into the network. The entire process of adjusting the neural network to input data elements and addition of nodes is repeated until the feature map satisfies requirements. A requirement is usually given in terms of a pre-determined minimum error value or a maximum number of allowed nodes.

It is a common practice in growing neural networks to locate new nodes between the node with the highest resource value and its connected nodes with high resource values. The resource value of a particular node in a neural network is a measure of the local neural network approximation error. In the DCS, the sum of distances between the weight vector of the node and the positions of all elements of the input data that have the node as its best matching unit. The goal behind dynamic allocation in the manner described above is to introduce new nodes into the regions within the neural network that represent high resource values, and thereby reduce the overall neural network approximation error. During our research with high-dimensional, multi-variate data from an adaptive flight control system, we observed the following undesirable conditions to occur during the process of dynamic allocation in neural networks.

- ***Undesirable Condition 1:*** The node with the highest resource value occasionally becomes completely disconnected from the neural network (UC1)
- ***Undesirable Condition 2:*** If two non-neighboring nodes become connected, then new nodes occasionally become located into regions in the neural network that do not represent any input data (UC2)

Both conditions mentioned above affect the robustness of the neural network algorithm and are undesirable for the realization of adaptive controllers using neural networks. Examples of neural networks that practise dynamic allocation with heavy dependance on resource values include, the Growing Cell Structures, the Growing Gas Model, the Growing Neural Gas, and the Dynamic Cell Structures (DCS) (7; 10; 11; 13). If the above mentioned conditions are not addressed, then adaptive controllers that use these neural networks are susceptible to unsafe and disastrous outcomes alike.

In this section, we analyze the learning properties of the neural network and the characteristics of the input data to understand the circumstances that lead to undesirable conditions during dynamic allocation in neural networks. We use the neural network learning algorithm from the DCS as a benchmark throughout this paper to demonstrate the undesirable conditions during dynamic allocation in neural networks. This is a justified approach given that the fundamental principles of dynamic allocation in most growing neural networks are related.

3.2 Analysis for UC1 (Undesirable Condition 1)

Dynamic allocation in neural networks is typically based on the highest resource value node and the structure of its connected neighboring nodes for placement of new nodes into the neural network map. According to the first undesirable condition (UC1) there is a chance that the highest resource value node can become completely disconnected from the neural network. Under such circumstance, the dynamic allocation algorithm will fail to locate new nodes into the neural network map causing errors during the execution of the algorithm.

Consider a high-dimensional data representative, $\mathcal{M} \subset \mathbb{R}^{\mathbb{D}}$ that is presented to a feature extracting neural network. The goal of the neural network is to generate a mapping, $\mathcal{G}_N : \mathcal{M} \to \mathbb{R}^{\mathbb{D}}$ that represents the information in $\mathcal{M}$ in a manner that retains its topology by using minimal number of network nodes. The representation, $\mathcal{G}_N \subset \mathbb{R}^{\mathbb{D}}$ implies the neural network map at a given time is made up of N nodes. The location of a node $i \in \{1, 2, \ldots, N\}$ in the neural network map is represented using weight vector, $w_i \in \mathcal{G}_N$. At a given time, two arbitrary nodes, $i, j \in \{1, 2, \ldots N\}$ in the neural network can establish a lateral connection. The strength of the connection between the two nodes is represented using a scalar value, $c_{ij} \in [0, 1]$. A connection strength of value 0 implies that the two nodes are disconnected and a value 1 represents two nodes becoming completely connected. The feature map, $\mathcal{G}_N$ is made up of node-locations and their connectivity structure, and can therefore be completely represented using weight vectors, w_i and connections, c_{ij}.

The resource value of a node, $i \in \{1, 2, \ldots, N\}$ will be represented as r_i, and its mathematical formulation is given below, where $m \in \mathcal{M} \subset \mathbb{R}^{\mathbb{D}}$ represents an element of the input data, $w_{bmu(m)}$ represents the weight vector of the best matching unit for m, and $d(m, w_{bmu(m)})$ represents the Euclidean distance between m and $w_{bmu(m)}$. A best matching unit for an input element is a node of the neural network that is closest to the location of the input element than any other node in the neural network.

$$r_i = \sum_{m \in \mathcal{M}} d(m, w_{bmu(m)}) \tag{1}$$

Consider 3 nodes, $i, j, k \in \{1, 2, \ldots, N\}$ of the neural network map, $\mathcal{G}_N$ at a certain time during the neural network's learning from the input data, $\mathcal{M}$. Let $m_{ij} \in \mathcal{M}$ represent an element of the input data that has the node, i as its best matching unit and the node, j as its second best unit. The second best unit is referred to the second closest node to the given input data element. Let $M_{jk} \subset \mathcal{M}$ represent the set of input data elements that have node j as their best matching unit and the node, k as their second best unit. Note that since the best matching unit of the input data element, m_{ij} is the node, i, it can be said that, $m_{ij} \notin M_{jk}$. Also, note that the situation described above is not hypothetical, but based on experimental observations from neural networks learning. Further details about the experiments will be provided in the Case Study section.

Without any loss of generality, let the node i remain as the highest resource value node in the neural for certain consecutive training cycles. The resource values of the node i be represented as, r_i. Also, let the weight vectors and the connectivity structure in the neural network are adjusted for the input data element m_{ij} before adjusting for the input data elements in the set, M_{jk}. Adjusting the connectivity structure of the neural network using the competitive Hebb rule for the input data element, m_{ij} will cause the nodes, i and j to establish a lateral connection with a strength of value 1. This is because the nodes i and j are the best matching unit and the second best unit respectively for the input data element, m_{ij}, and the competitive Hebb rule laterally connects the best matching and second best units with a connection strength of value 1, i.e. $c_{ij} = 1$ during this step. Without any loss of generality, consider the following conditions.

During the neural network learning to any element of the input data set M_{jk}, nodes j and k establish a lateral connection between them with a connection strength of value 1, $c(j, k) = 1$. As the competitive Hebb rule also decrements the connection strengths between the *bmu* and its connected nodes, the connection strength between the nodes i and j, $c(ij)$ that was earlier set to 1 will now become decremented by a factor of value β. The process of decrementing the connection strength between the nodes i and j will successively be repeated for all remaining elements of the set $\{m(j)\}$. If the set, $\{m(j)\}$ is large enough, then after a certain number of repetitions, the connection strength $c(ij)$ between the nodes i and j will fall below the pre-defined threshold, θ. In such a case, the competitive Hebb rule will disconnect the nodes i and j, $c(ij) = 0$. Due to last condition, the two nodes remain disconnected for all future adaptations.

Let n be defined as the number of times a lateral connection strength of value 1 be successively multiplied by the decrementing factor β in order for the connection strength to be equal to the threshold $\theta \in (0,1)$. We can compute n for a given θ and β to the following relation.

$$n = \frac{\log(\theta)}{\log(\beta)} \tag{2}$$

In our analysis, the number of elements in set $\{m(j)\}$ were assumed to be more than the value of n. This assumption is valid considering that n is a small finite number (usually about 4). In such a case, the node, i with the highest resource value becomes disconnected from every other node of the network for all future adaptations. According to the above analysis, if the dynamic allocation process were to determine the highest resource value node and its connected neighbor with a high resource value, the dynamic allocation algorithm will not be able to locate any new nodes into the neural networks. It is observed during our experiments with growing neural networks that the above described condition causes errors during the execution of the dynamic allocation algorithm.

3.3 Analysis for UC2 (Undesirable Condition 2)

During dynamic allocation in neural networks it is observed during our experiments with neural network learning that the highest resource value node can establish a lateral connection with nodes that are non-neighboring. By non-neighboring nodes here, we mean nodes in the neural network that map non-adjacent (distant) elements of the input data. According to the second undesirable condition (UC2), under these circumstances, new nodes can be placed into regions in the neural network that do not represent any input data. This is because dynamic allocation in neural networks between non-adjacent nodes is likely to introduce nodes into regions in the neural network that do not represent any data.

According to the definition of topology preservation in neural networks that is given in Sect. 1, a neural network mapping is not considered topology preserving if two non-neighboring nodes that represent non-adjacent input data elements establish a connection. By placing new nodes between non-neighboring nodes, the map generated by the neural network according to UC2 leads to a feature map that is not preserving the topology of the input data. The example provided in this section using the DCS algorithm demonstrates the actual circumstances of the neural network learning and the input data that lead the undesirable condition, UC2 to occur.

A neural network feature map, $\Phi : \mathcal{M} \subset \mathbb{R}^{\mathbb{D}} \rightarrow \mathcal{G}_N \subset \mathbb{R}^{\mathbb{D}}$ is generated by mapping the input data, $\mathcal{M}$ onto the neural network, $\mathcal{G}_N$, where N is the number of nodes in the neural network at a given time. An element of the input data, $m \in \mathcal{M}$ is mapped to a neural network node, $i \in \{1, 2, \ldots, N\}$ of the feature map, $\mathcal{G}_N$ such that the Voronoi polyhedron, $V_{bmu(m)}$ of the best matching unit for $m \in \mathcal{M}$ encloses m. Recall that for an input element, $m \in \mathcal{M}$, the best matching unit, $bmu(m)$ is the node of the neural network that is closest to m than any other node. The mapping Φ is considered neighborhood-preserving if the input data elements that are close in input data, $\mathcal{M}$ are mapped to adjacent nodes in the feature map, $\mathcal{G}_N$. The inverse mapping $\Phi^{-1} : \mathcal{G}_N \subset \mathbb{R}^{\mathbb{D}} \rightarrow \mathcal{M} \subset \mathbb{R}^{\mathbb{D}}$ is considered neighborhood-preserving if the weight vectors of adjacent nodes in $\mathcal{G}_N$ map input data elements that are close in $\mathcal{M}$. The feature map, $\mathcal{G}_N$ is said to have formed a topology preserving feature mapping of $\mathcal{M}$ if both mappings, Φ_w, and Φ_w^{-1} are neighborhood-preserving (12). It is important to realize from the definition of topology preservation given above that the lateral connection structure also plays an important role in the realization of a topology preserving feature mapping.

Based on the above presented definition of topology preservation, if it can be shown that during dynamic allocation either the mapping Φ_w, or Φ_w^{-1} is not neighborhood-preserving, then the feature map, $\mathcal{G}_N$ is not topology preserving with respect to $\mathcal{M}$. Consider that the neural network is presented with the data representative that is shown in Fig. 4. Also, consider that the neural network at a given time generates a feature map, $\mathcal{G}_{8A}$ as shown in Fig. 4. $\mathcal{G}_{8A}$ implies that the feature map generated by the neural network at the given time, A is made up of 8 nodes as shown in Fig. 4. The nodes in Fig. 4 are labeled 1 through 8 for the sake of convenience, and the numbering-order does not have any relevance here. Note that the feature map, $\mathcal{G}_{8A}$ shown in Fig. 4 is a topology preserving mapping of the input data $\mathcal{M}$ that is also shown in the background of Fig. 4. This is because at the given time both mappings, Φ_w, and Φ_w^{-1} are neighborhood-preserving.

Let node-1 be the best matching unit of an element of the input data, $m_1 \in \mathcal{M}$ that is highlighted in Fig. 4. Also, let node-8 be the second best unit for m_1. According to the competitive Hebb rule during updating the connectivity structure of the neural network for $m1$, nodes 1 and 8 establish a lateral connection between them with a connection strength of value 1. Fig. 5 shows this feature map generated by the neural network, where the map at time, B ($B > A$) is represented as $\mathcal{G}_{8B}$. Since nodes 1 and 8 represent non-neighboring nodes, a connection between them as shown in Fig. 5 implies that the inverse mapping, Φ_w^{-1} from $\mathcal{G}_{8B}$ to $\mathcal{M}$ is not neighborhood-preserving. According to the definition of topology preservation given earlier, the neural network mapping, $\mathcal{G}_{8B}$ is not considered as a topology-preserving feature map with respect to the input data $\mathcal{M}$ as the inverse mapping, Φ_w^{-1} is not neighborhood preserving. It can also be seen from Fig. 5 that node-2 is the only neighboring node for node-1.

Suppose that node-1 and node-8 remain connected during the dynamic allocation process with a non-zero connection strength, and let either of these nodes happen to be the node in the neural network with the highest resource value. For instance, let node 1 be the highest resource value node, and between node nodes 2 and 8 let the node 8 have a higher resource value. According to the current process of dynamic allocation in neural networks, a new node is introduced between node-1 and node-8 as shown in Fig. 6, where the new

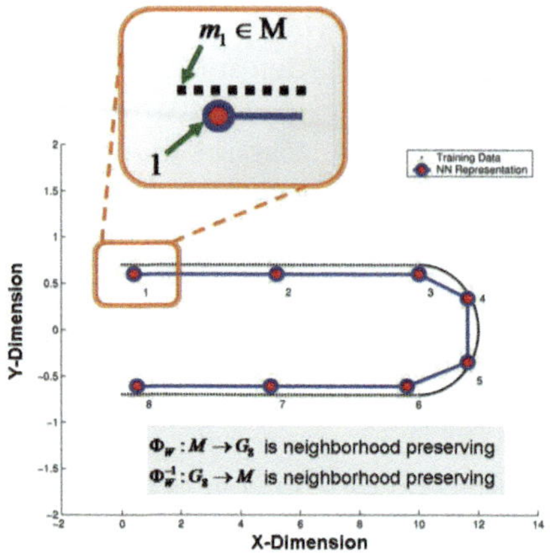

Fig. 4 Input data, $\mathcal{M}$ and the neural network feature map, $\mathcal{G}_{8A}$ using 8 nodes at time, A

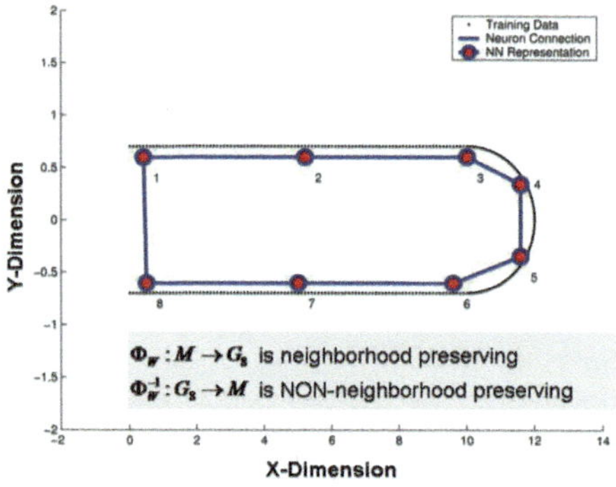

Fig. 5 Input data, $\mathcal{M}$ and the neural network feature map, $\mathcal{G}_{8B}$ using 8 nodes at time, B $(B > A)$

map is represented by, $\mathcal{G}_{9C}$. Note that by introducing a new node into a region that does not represent any data (useful information) the dynamic allocation process further diminishes the original topology of the feature map, $\mathcal{G}_{8B}$ with respect to the input data, $\mathcal{M}$.

In the above described example both, node-2 and node-8 are connected to the node-1, only node-2 is the neighbor for node 8. However, since node-8 has a resource value that is higher than that of node-2, the current process of dynamic allocation assumes that the node-8 is a neighbor for node-1. Thereby, it introduces a new node between nodes 1 and 8. It is observed during this research that such conditions occur frequently during neural networks adaptation to a non-stationary data. The reason being that current dynamic allocation processes blindly introduce nodes without realizing if the highest resource node and its connected node with the highest resource value are actually neighbors. It is also obvious from the above discussions that current

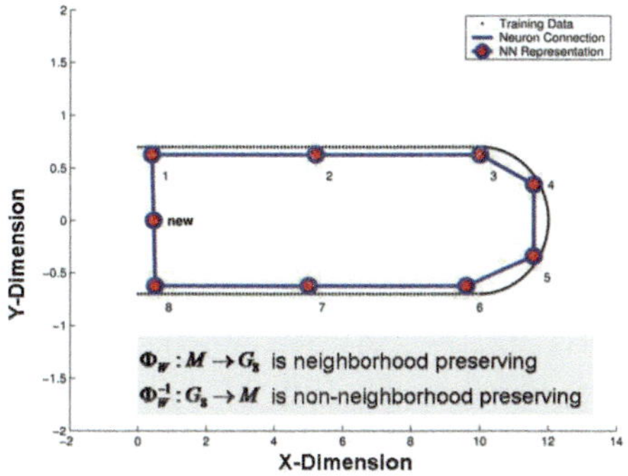

Fig. 6 Input data, $\mathcal{M}$ and the neural network feature map, $\mathcal{G}_{9C}$ using 9 nodes at time, C

dynamic allocation processes assume that under all conditions of data, the node that has the highest resource value of all nodes connected to the node in the neural network with the highest resource value are neighbors. The idea behind this assumption is that if two nodes are connected, and have high resource values, then they are likely neighbors.

In our research we found that the assumption is not always valid, and is only valid if the feature map generated by the neural network is topology preserving with respect to the given input data. For example, the assumption while valid for the feature map, $\mathcal{G}_{8A}$ that is shown in Fig. 4, becomes invalid for the feature map, $\mathcal{G}_{8B}$ that is shown in Fig. 5. Because in $\mathcal{G}_{8B}$ two nodes that are non-neighboring establish a connection between them, the inverse mapping, Φ_w^{-1} becomes non-neighborhood-preserving. It is, therefore, not valid to assume that two connected nodes are adjacent, unless the feature map is topology-preserving. Note that a topology-preserving mapping by default implies that the mappings, Φ_w and Φ_w^{-1} are neighborhood-preserving.

If the above described undesirable condition during dynamic allocation needs to be prevented, then it is necessary to ensure that one of the following two conditions is always satisfied.

1. A new node is always located into a region between two neighboring nodes
2. A connection can be established between two nodes only if the two nodes are neighbors

To satisfy either of these conditions, it is necessary to identify if any two given nodes are neighbors. The question then becomes: how can we identify neighboring nodes in a neural network? Identification of neighboring nodes in the inverse mapping, Φ_w^{-1} may require significant computation. First, the input data elements need to be grouped into sets, $M_1, M_2, \ldots, M_N$, where N is the number of nodes in the neural network. The grouping should be performed such that all input elements of the set M_i, have the node $i \in \{1, 2, \ldots, N\}$ as their best matching unit. Note that, $M_1 \cup M_2 \cup \ldots \cup M_N = M \in \mathbb{R}^{\mathbb{D}}$ since each input element can only have a single node as its best matching unit.

In the above example, the process would lead to grouping M into sets, $M_1, M_2, \ldots, M_8$. Next, the distance between each element of the input data set, M_i and each elements of all other input data sets, M_j are computed, where $j \in \{1, 2, \ldots, N\}$ and $j \neq i$. Let the distance sets be represented as, $\{d(M_i, M_j)\}$. A node j can be said to be a neighbor for node i if the minimum value of the set, $\{d(M_i, M_j)\}$ is obtained from an input element of the set M_j. In the above example, the distance sets for node-1 can be represented as, $\{d(M_1, M_2)\}, \{d(M_1, M_3)\}, \ldots, \{d(M_1, M_8)\}$. For the above example, the minimum value of this sets will come from an element in the set, $\{d(M_1, M_2)\}$. Thereby, it is node-2 (and not node-8) that is actually a neighbor for node-1.

It is evident from the above discussion that determining neighboring nodes in a neural network in the manner described above requires significant computational resources. In order to ensure that only neighboring nodes establish a connection between them as stated in the 2nd condition one would need to perform the above described computations for determining neighboring nodes every time any weight vector in the neural network is adjusted. Because the neural network adjusts its weight vectors every time an element of the input data is presented to it, the 2nd condition will require performing the above described computations for determining neighboring nodes for each input data element in a single learning cycle. Also, it is shown in the work of (12) that neural networks that use competitive Hebb rule to update connections between nodes in the long run are able to form a topology preserving mapping. In essence, the work of (12) shows that is not necessary to determine the neighboring nodes in a neural network that use the competitive Hebb rule. Our work is focused on the 1st condition, i.e. to ensure that a new node is always located into a region between two neighboring nodes. Moreover, since a new node is located after the neural network adjusts its weight vectors and connections for all input data element, satisfying the 1st condition requires relatively less computational overhead than satisfying 2nd condition.

The map generated by the neural network at time C is not topology preserving due to the location of the new node. The initial location of the new node at time C is shown in Fig. 7 as new-initial. However, as the competitive Hebb rule in the long run, generates a topology preserving mapping ((12)) a new node that is introduced between non-neighboring nodes is eventually absorbed into a connection between two neighboring nodes. The final location of the new node at time D is shown in Fig. 7 as new-final. The map due to the final location of the new node is topology preserving.

It was observed during our experiments with neural networks learning from high-dimensional data in an adaptive control system that $D \gg C$, i.e. the neural network takes significant learning iterations (between the order of 100 to 300) to correct the feature map from a non-neighboring to a topology preserving mapping. More learning iterations for correcting the mapping implies more computational overhead and a delay in the time for the formation of a topology-preserving feature map. During the time it takes the neural

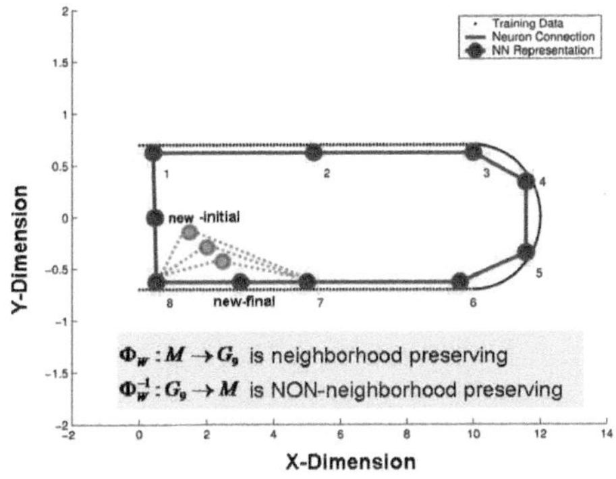

Fig. 7 Initial and final locations of the Neural network feature map using 9 nodes at time, D ($D \gg C$)

network to correct the map, if the adaptive controller were to use the feature map generated by the neural network, then it can lead to an undesirable controller output. This is especially true given that no single metric can be used to completely determine the accuracy of topology preserved by the feature map.

In order to overcome the problems presented in this section during dynamic allocation in neural networks a modified dynamic allocation process is presented in the following section. The details of the algorithm will be discussed in the following section.

4 Data-Driven Dynamic Allocation Algorithm

The process of dynamic allocation in neural networks commonly relies on the location and the connectivity structure of the highest resource value nodes for placement of new nodes into the neural network. It is assumed during dynamic allocation in neural networks that nodes connected to the highest resource value node are neighboring both, in the forward and inverse mappings, Φ_w and Φ_w^{-1}. In the previous section, it is shown that this assumption is not always valid as nodes connected to the highest resource value node may not necessarily be neighboring in the inverse mapping Φ_w^{-1}. It was demonstrated in the previous section using examples from neural network learning that the feature mapping resulting from this approach of dynamic allocation is not fully robust in terms of forming a reliable topology preserving mapping from the input data (UC2). It is also shown that dynamic allocation in neural networks in this manner can become error-prone when the highest resource value node becomes disconnected from all other nodes in the neural network (UC1). To address these issues, we present a dynamic allocation algorithm for growing neural networks that is driven by the distribution of the input data. The advantage of the algorithm is that it introduces new nodes into locations in the neural network that strictly represent input data and does not fully rely on the connectivity structure of the highest resource value node. The major difference in this dynamic allocation algorithm is that is driven by the distribution of the input data rather than the connectivity of the resource information.

Following are the steps that describe the proposed dynamic allocation algorithm.

Step-1: *Find the node of the neural network that has the highest resource value as given in Equation-1.* Let $r_{hrn} \in \mathbb{R}$ represent the highest resource value, and $w_{hrn} \in \mathcal{G}_N \in \mathbb{R}^D$ represent the weight vector of the highest resource value node, where $\mathcal{G}_N$ is the feature map of the neural network with N nodes.

Step-2: *Sort the input data elements into N sets by grouping the input elements that share a given node as their best matching units as one set.* Let $M_1, M_2, \ldots, M_N \in \mathbb{R}^D$ represent these groups that have the nodes, $1, 2, \ldots, N$ as their best matching units. Note that $M_1 \cup M_2 \cup \ldots \cup M_N = \mathcal{M}$, where $\mathcal{M} \in \mathbb{R}^D$ is the input data. Let M_{hrn} be the set of input elements that have the highest resource value node as their best matching unit. Let $\#M_{hrn}$ represent the number of elements in the set, M_{hrn}.

Step-3: *Locate a new node into the neural network by placing it over the input data element of M_{hrn} that has the longest distance from w_{hrn}.* If $w_{new} \in \mathcal{G} \in \mathbb{R}^D$ represents the weight vector of the new node in the feature map $\mathcal{G}$, then $w_{new} = m^*$, where m^* is determined by, $max_{m^* \in M_{hrn}} \{d(m^*, w_{hrn})\}$.

Step-4: *Update the lateral connections between the nodes using the competitive Hebb rule for the input data elements in the set, M_{hrn}.* Refer to the work of (8; 12) for detailed description of the competitive Hebb rule.

Step-5: *Determine the set of nodes, N_{cnew} that become connected to the new node.* Let $\#N_{cnew}$ represent the total number of nodes that become connected to the new node. Note that $\#N_{cnew} \geq 1$ as at least one node (the second best unit for input element, m^*) becomes connected to the new node because $m^* = w_{new}$.

Step-6: *Apply Step-1 again to sort the input data elements into groups based on their best matching units. Determine the sets of input elements $M_i \subset \mathcal{M}$ that have nodes, $i \in N_{cnew}$ as their best matching unit.*

Step-7: *Adjust the resource values for the new node and the nodes connected to the new node as given below.*

$$r_{new} = \frac{1}{(\#M_{hrn})(\#N_{cnew})} \sum_{i \in N_{cnew}} (\#M_i)(r_i)$$

$$r_i = r_i - \frac{(\#M_i)}{(\#M_{hrn})(\#N_{cnew})}(r_i) \qquad \forall i \in N_{cnew}$$

As described in the previous section, it is possible for the highest resource value node to become completely disconnected from the neural network (UC1). Traditional approaches of dynamic allocation search for nodes connected to the highest resource value node for placement of new nodes into the neural network (7; 8; 10; 13; 15). This is an undesirable condition that will cause execution-errors during the operation of the dynamic allocation algorithm. The above proposed dynamic allocation algorithm overcomes this problem by not relying on the connectivity structures of the highest resource value node for placement of new nodes into the neural network. Instead, the proposed dynamic allocation algorithm relies on the locations of the input data elements that have the highest resource value as their best matching units for placement of new nodes into the neural network. Step-2 of the above presented dynamic allocation algorithm describes the details of this process.

Current approaches of dynamic allocation are based on placement of new nodes into the neural network on the connection between the highest resource value node and one of its connected node (7; 8; 10; 13; 15). It is demonstrated in the previous section that the highest resource value node can establish a lateral connection to a non-neighboring node (UC2). It was demonstrated in the previous section that it is possible for dynamic allocation algorithms to place new nodes into the neural network between non-neighboring nodes. The region connecting non-neighboring nodes is likely not representing any input data. Addition of new nodes into regions of the neural network that do not represent any useful information is an undesirable condition for dynamic allocation that will lead to the formation of a feature map that is not preserving the topology of the input data. This is an undesirable condition that will cause execution-errors during the operation of the dynamic allocation algorithm. The above proposed dynamic allocation algorithm overcomes this problem by not relying on the connectivity structures of the highest resource value node for placement of new nodes into the neural network. Instead, the proposed dynamic allocation algorithm locates new nodes into the neural network by placing them over the location of the input data elements that are not properly represented by the highest resource value node. Step-3 of the above presented dynamic allocation algorithm describes the details of this process.

Realize that resource value represents a local error measure for the neural network (7; 8; 10; 13; 15). If a node has a high resource value in a neural network, then it implies that the node is not representing the information from the set of input data elements that have the node as their best matching units. By placing new nodes over an input data element from this set, ensures that the new nodes is a neighbor for the highest resource value node. Also placing new nodes over the most poorly represented input data element from this set will likely increase the rate of change of the accuracy for the overall neural network mapping. Quantitative details about this behavior will be discussed in the following section.

The new node establishes a lateral connection with at least one other node of the neural network because the lateral connections between all nodes are refreshed during Step-4 of the above proposed dynamic allocation algorithm using the competitive Hebb rule. Also, the feature map will retain its neighborhood-preserving property during Step-4 because of the manner in which new nodes are located into the neural network during Step-2 and 3 of the proposed dynamic allocation algorithm.

Conventionally, dynamic allocation algorithms assign a resource value to the new node based on the resource value of the highest resource value node and the resource values of its connected nodes. The resource values of the nodes connected to the new node are also distributed based on their original resource values (7; 8; 10; 13; 15). It is demonstrated in the previous section that it is possible for the highest resource value node to establish a lateral connection with non-neighboring nodes (UC2). Given the possibility of the condition, UC2 it is not proper to follow the conventional approach of assigning resource value to the new node or distribution of the resource values of the nodes connected to the new node. The proposed dynamic allocation algorithm updates the resource values of the new node and its connected nodes based on distribution of the input data elements surrounding these nodes. This process is described in the proposed dynamic allocation in Steps 5, 6, and 7. The steps described in the proposed dynamic allocation algorithm ensure that placement of new nodes into a neural network, assignment of resource value for the new nodes, and distribution of resource values of the nodes connected to the new node are all driven by the distribution of the input data surrounding the highest resource value node.

The following analysis will shown that the neural network approximation error continually decreases for each iteration for a dynamic allocation process that add new nodes into the neural network by overlaying them on input data elements. The analysis will also show that this is not the case for a dynamic allocation process that is based on addition of new nodes into regions in the neural network between high resource values nodes. The resource value of the masked Voronoi regions of a neural network, which represents a quantitative metric for measuring the neural network's approximation error (see ()) will be used as a benchmark for the analysis. The following equation provides the mathematical formulation for the resource values of the masked Voronoi region, where V is the volume of the masked Voronoi region.

$$R_V = \sum_{m \in V} ||m - w|| \tag{3}$$

A dynamic allocation process can add new nodes between the highest resource value nodes into a region that does not represent any input data as demonstrated in the example shown in 5. In such a case, the new node is not the best matching unit for any input data element. The resource value of the masked Voronoi regions immediately after the addition of the new node, $R_{V_{\text{after}}}$ is not less than the value before the addition, $R_{V_{\text{after}}}$.

$$R_{V_{after}} \nless R_{V_{bef}} \tag{4}$$

If the newly introduced node is overlaid over an input data element then the new node becomes the best matching unit for that input data element. In other words, the new node becomes the best matching unit for at least one input data element. The Euclidean distance between the newly introduced node and the input data element over which it is overlaid is zero, which will be value that is lower than the Euclidean distance between the same input data pattern and its previous best matching unit. Therefore, the resource value of the masked Voronoi region after the addition of a new node by overlaying over an input data element will always become smaller than the resource value of the masked Voronoi region before the addition,

$$R_{V_{\text{after}}} < R_{V_{\text{before}}} \tag{5}$$

The above result shows that a dynamic allocation process based on addition of new nodes over input data elements will continue to decrease each iteration. The significance of the proposed dynamic allocation algorithm in the context of adaptive controller applications is that neural networks can have the ability to generate robust topology-preserving feature mappings of the non-stationary input data. This will be demonstrated in the following section using data from an experimental adaptive flight control simulator (23) that uses neural network to account for variations in the aircraft system parameters in real-time during the flight.

5 Case Study

In the recent years, NASA has conducted series of experiments evaluating neural networks for providing fault-tolerant flight control capabilities following sensors and actuators failures (14; 21; 22). The results from these experiments have been used in the development of the NASA's, Intelligent Flight Control System, the IFCS. The system was primarily designed to incorporate online learning neural network into flight control to improve aircraft performance during no-failure flight conditions, and to optimize the performance during a single or multiple failure conditions.

Fig. 8(a) shows the architectural overview of the first generation of the IFCS implementation. The first generation of the IFCS implementation consists of four major components. The first component is the Pre-Trained Neural Networks, the PTNN, which are used to produce stability and control derivatives for specific flight conditions. It is previously trained and is fixed during the flight. Secondly, there is a control algorithm that uses Stochastic Optimal Feedforward and Feedback Technique, SOFFT to provide optimal stability and specific flying characteristics. The third component is the Parameter Identifier, the PID that determines the stability and control characteristics of the aircraft in real-time. The fourth component is the Online Learning Neural Network, the OLNN that learns the difference in the values of the PTNN and PID during the flight. The difference in the values of the PTNN and the PID indicate conditions falling outside the scope of traditional or linearized control gain look-up tables. The role of the OLNN is tostar represent and store the SCD error values in-flight, and provide an estimate of the stored SCD errors for control under new flight conditions. The Dynamic Cell Structure, the DCS neural networks is used in implementing the OLNN in the first generation of the IFCS.

For our analysis, we use the control data obtained from the NASA-WVU high fidelity, flight simulator that is located in the Mechanical and Aerospace Department at West Virginia University (22; 23). The simulator is built over the architecture and the functionality of the first generation IFCS. The user interface of the flight-simulator is shown in Fig. 8(b).

The control data collected from the simulator consists of 4 system parameters, (1) aircraft altitude, (2) velocity, (3) angle of attack, α, and (4) sideslip angle, β, and 2 control derivatives, (a) partial derivatives of the normal force with respect to the angle of attack, $Cz\alpha l$, and (b) partial derivatives of the normal force with respect to the stabilator deflection, $Cz\alpha s$. Data is collected from the flight simulator for nine flight-conditions, including two nominal or no-failure, two surface failure, and five control failure flight conditions. The flight conditions for each case are listed in detail in Table 1.

Two on-line learning algorithms are implemented using MATLAB software, the first uses the original DCS dynamic allocation approach ((8; 14)), and the second uses the proposed dynamic allocation approach (the algorithm is provided in the previous section). The six dimensional control data obtained from the flight simulator is applied to the two algorithms to compare their performance in adaptation. The OLNN introduces a single node into the neural network at the end of each learning cycle. We use the locations of the newly introduced nodes from the two algorithms for comparison. The OLNN extracts features from the control data in the input space, and represents the information in the same dimensional feature space, i.e. both, the inputs and outputs of the OLNN are six-dimensional.

Fig. 9 shows the six dimensional control data (in blue) and the locations of the new nodes (in red) for a no-failure flight condition. Fig. 9(a), 9(c), and 9(e) shows the control data and the placement of new nodes using the original dynamic allocation approach for dimensions, 1 Vs 2, 3 Vs 4, and 5 Vs 6 respectively. Fig. 9(b), 9(d), and 9(f) show the corresponding data using the proposed dynamic allocation approach for dimensions, 1 Vs 2, 3 Vs 4, and 5 Vs 6 respectively. The plots show that the proposed dynamic allocation approach always adds new nodes into the neural network over the

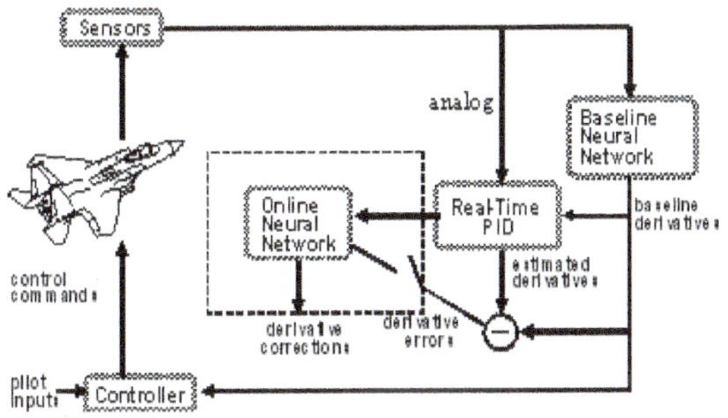

(a) Architectural overview of NASA's first generation IFCS

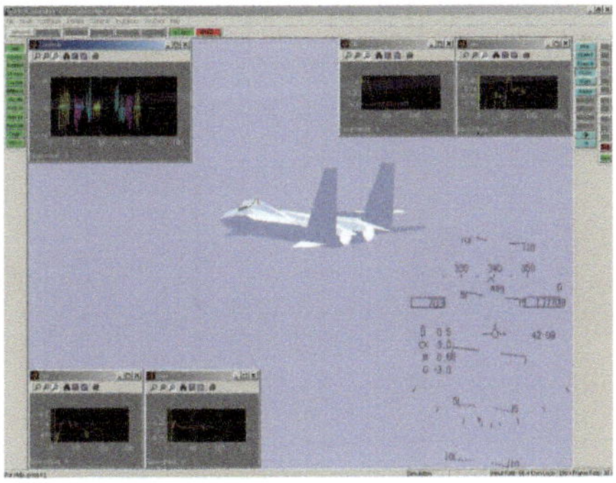

(b) User interface of the high fidelity NASA-WVU flight simulator

Fig. 8 Architectural overview of NASA's first generation IFCS and the user interface of the high fidelity NASA-WVU flight simulator that is built over the first generation IFCS architecture

Table 1 Tested conditions of adaptive control system operation

	Failure Type	Failure Description
1	No-Failure Mode 1	–
2	No-Failure Mode 2	–
3	Surface-Failure Mode 1	50% loss of left stabilator
4	Surface-Failure Mode 2	50% loss of right aileron
5	Control-Failure Mode 1	Left stabilator locked at 0 deg
6	Control-Failure Mode 2	Left stabilator locked at +3 deg
7	Control-Failure Mode 3	Left stabilator locked at -3 deg
8	Control-Failure Mode 4	Right aileron locked at +3 deg
9	Control-Failure Mode 5	Right aileron locked at -3 deg

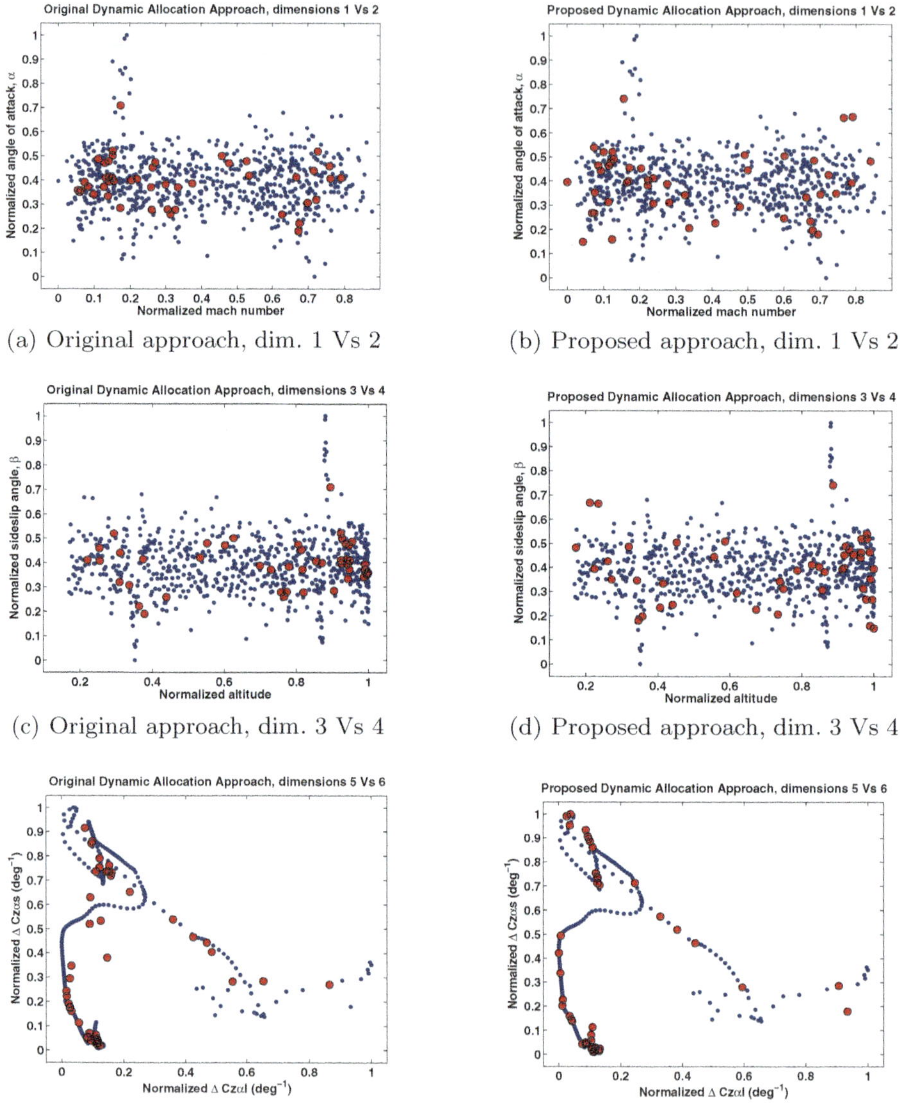

(a) Original approach, dim. 1 Vs 2

(b) Proposed approach, dim. 1 Vs 2

(c) Original approach, dim. 3 Vs 4

(d) Proposed approach, dim. 3 Vs 4

(e) Original approach, dim. 5 Vs 6

(f) Proposed approach, dim. 5 Vs 6

Fig. 9 Comparison of the original and the proposed dynamic allocation approaches. New nodes (red) and control data (blue) for a no-failure flight condition

control data. Whereas, the original approach occasionally adds new nodes into regions in the neural network that does not represent data, causing the new nodes to become inactive (these nodes are called dead-nodes). Notice that the issue with dead nodes occurs even during nominal flight conditions if the adaptation is performed using the original dynamic allocation approach

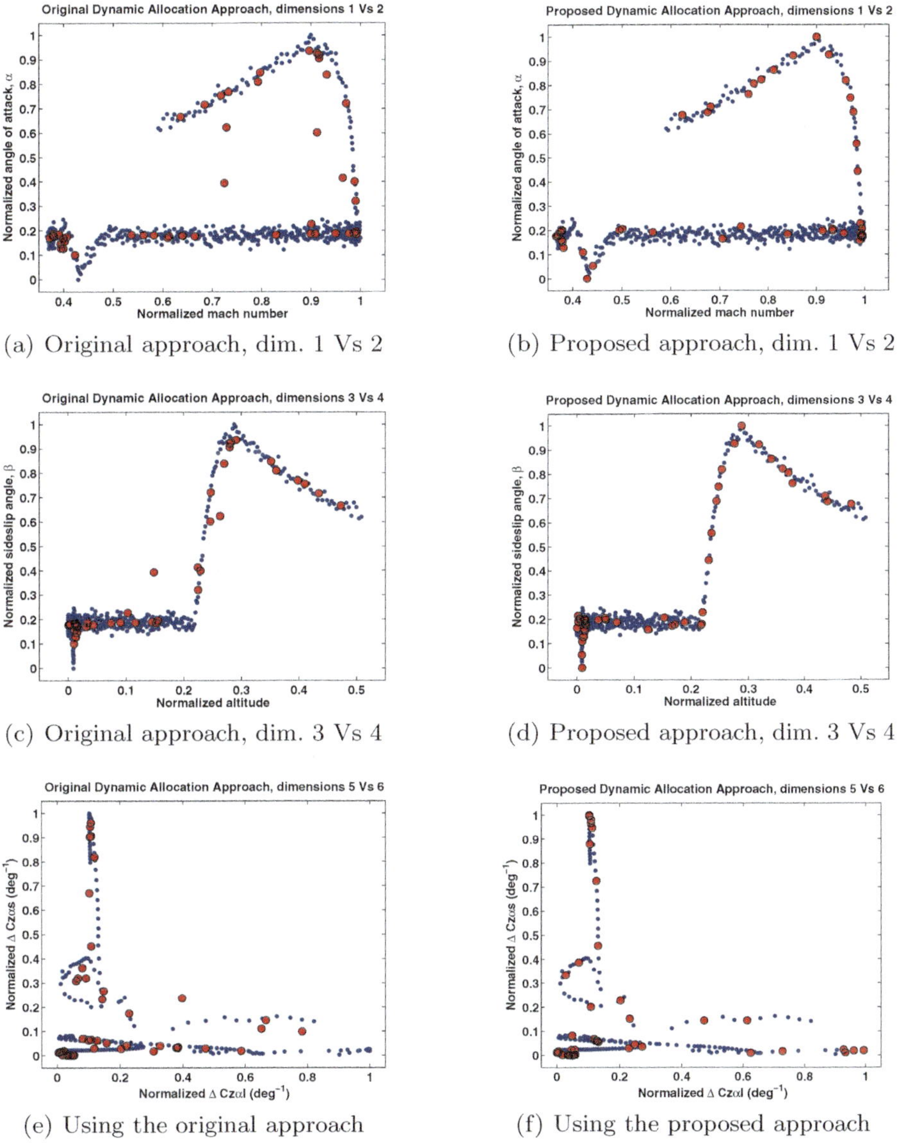

(a) Original approach, dim. 1 Vs 2

(b) Proposed approach, dim. 1 Vs 2

(c) Original approach, dim. 3 Vs 4

(d) Proposed approach, dim. 3 Vs 4

(e) Using the original approach

(f) Using the proposed approach

Fig. 10 Comparison of the original and the proposed dynamic allocation approaches. New nodes (red) and control data (blue) for a surface-failure flight condition

as demonstrated in this case. It can also be seen from the plots that the placement of new nodes using the proposed approach generates a mapping that accurately captures the topology of the control derivative data, i.e the new nodes are over the data and are well distributed. This is because the

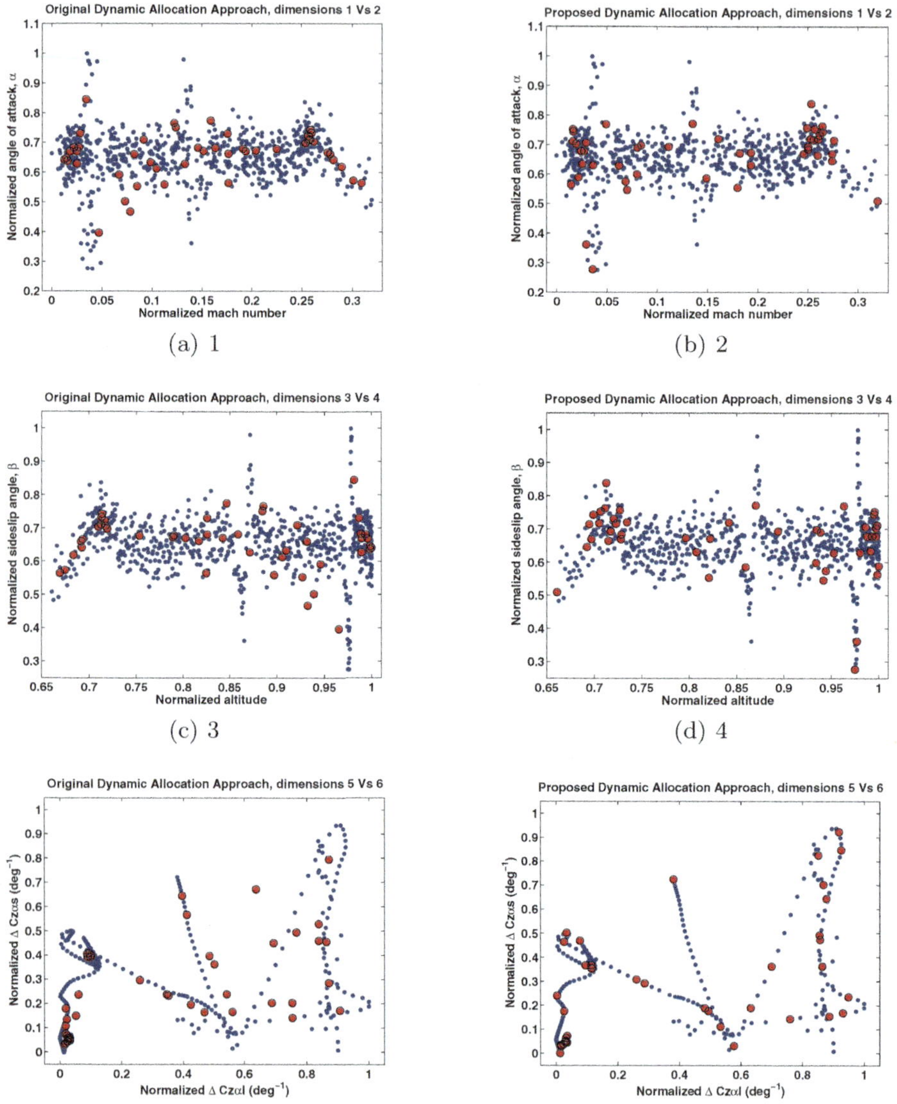

(a) 1 (b) 2

(c) 3 (d) 4

(e) Using the original approach (f) Using the proposed approach

Fig. 11 Comparison of the original and the proposed dynamic allocation approaches. New nodes (red) and control data (blue) for a control-failure flight condition

proposed approach adds new nodes into regions in the the network over the control data, where there is a necessity for new nodes.

Fig. 10 shows the control derivative data (in blue) and the locations of the new nodes (in red) for a surface-failure flight condition. Fig. 10(a), 10(c), and 10(e) shows the control data and the placement of new nodes using the original dynamic allocation approach for dimensions, 1 Vs 2, 3 Vs 4, and 5 Vs 6 respectively. Fig. 10(b), 10(d), and 10(f) show the corresponding data using the proposed dynamic allocation approach for dimensions, 1 Vs 2, 3 Vs 4, and 5 Vs 6 respectively. The plots show that even in the case of a surface-failure condition, the proposed dynamic allocation approach always adds new nodes into the neural network over control data. Whereas, the original approach occasionally adds new nodes into regions in the neural network that does not represent data, causing them to become dead-nodes. It is important to notice that even during surface-failure flight conditions, the placement of new nodes using the proposed approach generates a mapping that accurately captures the topology of the control derivative data, i.e the new nodes are over the data and are well distributed. This result demonstrates that the proposed dynamic allocation approach has the ability to perform not only in nominal flight conditions, but also during the system's adaptation from a failed surface flight condition.

Fig. 11 shows the control derivative data (in blue) and the locations of the new nodes (in red) for a control-failure flight condition. Fig. 11(a), 11(c), and 11(e) shows the control data and the placement of new nodes using the original dynamic allocation approach for dimensions, 1 Vs 2, 3 Vs 4, and 5 Vs 6 respectively. Fig. 11(b), 11(d), and 11(f) show the corresponding data using the proposed dynamic allocation approach for dimensions, 1 Vs 2, 3 Vs 4, and 5 Vs 6 respectively. The plots show that even in the case of a control-failure condition, the proposed dynamic allocation approach always adds new nodes into the neural network over control data. Whereas, the original approach occasionally adds new nodes into regions in the neural network that does not represent data, causing them to become dead-nodes. Even during control-failure flight conditions, the placement of new nodes using the proposed approach generates a mapping that accurately captures the topology of the control derivative data, i.e the new nodes are over the data and are well distributed. The results demonstrate that the proposed dynamic allocation approach has the ability to perform in nominal, and also during the system's adaptation from a failed surface and failed control conditions.

The control data obtained from the NASA-WVU adaptive flight simulator is applied to the original and the proposed dynamic allocation algorithms. The results presented in this section practically demonstrate that the proposed dynamic allocation approach well distributes the neural network nodes over the control data than the original approach. The results also show that the proposed approach always adds new nodes over the data, and thereby produces a more accurate representations of the data than the original approach. It is also demonstrated that the ability of the proposed approach to overlay new nodes on the control data allows it to overcome the issue with dead-nodes.

6 Conclusion

A major challenge in the application of self-organizing neural networks in adaptive controllers is the development of a reliable and robust dynamic allocation approach, i.e. a process of adding new nodes to the neural network during adaptation that efficiently and effectively captures the topology of the varying control data. In this paper, we have discussed some of the issues that exist in current approaches of dynamic allocation that are undesirable for adaptive control applications. In order to overcome these issues, a novel dynamic allocation approach is introduced to strategically add new nodes over the input data into the regions in the neural network where there is a necessity for new nodes. Using the multivariate control data obtained from a flight simulator for no-failure, surface-failure, and control-failure flight conditions, it is demonstrated that the proposed approach of dynamic allocation generates mapping that are well distributed and accurately capture the topology of the control data, as well as overcomes the common problems that are encountered in adaptation of a neural network to non-stationary data.

The study and development of efficient and effective dynamic allocation approaches in neural networks, like the one presented in this paper are challenging, yet important to the design and implementation of reliable and high performance adaptive controllers. The dynamic allocation approach presented in this paper is not necessarily the ultimate solution for on-line adaptation, but is only one of the several methods of strategic on-line adaption for controller neural networks.

References

1. van der Putten, P.: Utilizing the Topology Preserving Property of Self-Organizing Maps for Classification. M.S. Thesis. Cognitive Artificial Intelligence, Utrecht University, NL (1996)
2. CMU Benchmark Archive (2002), `http://www-2.cs.cmu.edu/afs/cs/project/ai-repository/ai/areas/neural/bench/cmu/`
3. Fahlman, S.E.: CMU Benchmark Collection for Neural Net Learning Algorithms, Carnegie Mellon Univ, School of Computer Science, Machine-readable Data Repository, Pittsburgh (1993)
4. Perhinschi, M.G., Campa, G., Napolitano, M.R., Lando, M., Massotti, L., Fravolini, M.L.: A Simulation Tool for On-Line Real Time Parameter Identification. In: Proc. of the 2002 AIAA Modeling and Simulation Conference, Monterey (August 2002)
5. Fritzke, B.: Growing Self-organizing Networks - why? In: European Symposium on Artificial Neural Network, pp. 61–72 (1996)
6. Ahrns, I., Bruske, J., Sommer, G.: On-line Learning with Dynamic Cell Structures. In: Proc. of the International Conference on Artificial Neural Networks, vol. 2, pp. 141–146 (1995)

7. Bruske, J., Sommer, G.: Dynamic Cell Structures. In: Neural Information Processing Systems (NIPS1995), vol. 7, pp. 497–504 (1995)
8. Bruske, J., Sommer, G.: Dynamic Cell Structure Learns Perfectly Topology Preserving Map. Neural Computations 7(4), 845–865 (1995)
9. Kohonen, T.: The Self-Organizing Map. Proc. of the IEEE 78(9), 1464–1480 (1990)
10. Fritzke, B.: A Growing Neural Gas Network Learns Topologies. In: Advances in Neural Information Processing Systems, vol. 7, pp. 625–632. MIT Press, Cambridge (1995)
11. Fritzke, B.: Growing Grid - A Self-Organizing Network With Constant Neighborhood Range and Adaptation Strength. Neural Processing Letters 2(5), 9–13 (1995)
12. Martinetz, T., Schulten, K.: Topology Representing Networks. Neural Networks 7(3), 507–522 (1994)
13. Fritzke, B.: Growing Cell Structures - A Self-organizing Network for Unsupervised and Supervised Learning. ICSI 7(9), 1441–1460 (1993)
14. Jorgensen, C.: Feedback Linearized Aircraft Control Using Dynamic Cell Structures. In: World Automation Congress, ISSCI 050.1-050.6, Alaska (1991)
15. Fritzke, B.: Unsupervised Clustering With Growing Cell Structurea. In: Proc. of the IJCNN, pp. 531–536 (1991)
16. Kohonen, T.: The Self-Organizing Map. Proc. of the IEEE 78(9), 1464–1480 (1990)
17. Farkas, I., Chud'y, L.: Modified Dynamic Cell Structures as a Thinning Algorithm. In: Sinc'ak, P. (ed.) Proc. of 1st Slovak Neural Network Symposium, November 1996, pp. 71–80 (1996)
18. Ahrns, I., Bruske, J., Sommer, G.: On-Line Learning with Dynamic Cell Structure. In: Proc. of International Conference on Artificial Neural Networks, vol. 2, pp. 141–146 (1995)
19. Bruske, J., Riehn, L., Hansen, M., Sommer, G.: Dynamic Cell Structures for Calibration–free Adaptive saccade control of a four-degrees-of-freedom binocular head. Technical Report TR-9608, Institut f. Informatik und Praktische Mathematik (1996)
20. Herpers, R., Witta, L., Bruske, J., Sommer, G.: Dynamic Cell Structures for the Evaluation of Keypoints in Facial Images. International Journal of Neural Systems 8(1), 27–39 (1997)
21. Burcham Jr., F.W., Maine, T.A., Fullerton, C.G., Webb, L.D.: Development and flight evaluation of an emergency digital flight control system using only engine thrust on an F-15 airplane. Technical Report TR-3627, NASA (1996)
22. Napolitano, M., Neppach, C.D., Casdorph, V., Naylor, S., Innocenti, M., Silvestri, G.: A neural network-based scheme for sensor failure detection, identification, and accommodation. AIAA Journal of Control and Dynamics 18(6), 1280–1286 (1995)
23. Napolitano, M., Molinaro, G., Innocenti, M., Martinelli, D.: A complete hardware package for a fault tolerant flight control system using on-line learning neural networks. IEEE Control Systems Technology (1998)

Immune Systems Inspired Approach to Anomaly Detection, Fault Localization and Diagnosis in Automotive Engines

Dragan Djurdjanovic, Jianbo Liu, Kenneth A. Marko, and Jun Ni

Abstract. As more electronic devices are integrated into automobiles to improve the reliability, drivability and maintainability, automotive diagnosis becomes increasingly difficult. Unavoidable design defects, quality variations in the production process as well as different usage patterns make it is infeasible to foresee all possible faults that may occur to the vehicle. As a result, many systems rely on limited diagnostic coverage provided by a diagnostic strategy which tests only for a priori known or anticipated failures, and presumes the system is operating normally if the full set of tests is passed. To circumvent these difficulties and provide a more complete coverage for detection of any fault, a new paradigm for design of automotive diagnostic systems is needed. An approach inspired by the functionalities and characteristics of natural immune system is presented and discussed in the paper. The feasibility of the newly proposed paradigm is also demonstrated through application examples.

1 Introduction

The immune system of a living organism serves the purpose of defending the organism from external intrusions, such as microbes, viruses, or foreign bodies. Natural immune systems accomplish anomaly detection, identification and isolation of the root cause of anomalous behavior as well as elimination of the problem. Improper function or failure of the immune system causes serious problems in the organism and frequently leads to its death (illnesses such as leukemia, cancer or AIDS manifest themselves through failures of the immune system to perform its functions). In the human society, many thousands of years of the evolution and development of civilization are marked by dramatic improvements in tools and products that are used. Breakthrough inventions and iterative design improvements result in products that have more power, deliver more functionality, are easier to manufacture and use

J. Schumann, Y. Liu (Eds.): Appl. of Neural Networks in High Assur. Sys., SCI 268, pp. 141–163.
springerlink.com

etc. This process can be seen as being analogous to the process of natural evolution, in which living organisms have passed through series of "design iterations" of whom the strongest, most adaptive and most functional have progressed to yield even more powerful and functional natural designs.

In man's creations, engineering designs have been equipped with increasingly sophisticated capabilities for system performance diagnosis and prediction, especially in the case of sophisticated, expensive and safety critical systems, such as manufacturing equipment (1; 2; 3), computer networks (4), (5), automotive systems (6), (7), chemical plants (8), (9), etc. The purpose of these capabilities is to ensure proper functionality of the system and can be seen as analogous to the role of the immune system in a living organism.

Nevertheless, the way current diagnostic and prognostic capabilities are realized in automotive engines (and most other applications for that matter) is significantly different from the way the immune system in a living organism is realized. Condition monitoring systems are based on sophisticated feature extraction, performance assessment, performance prediction and training algorithms that are centered around recognizing and quantifying indications of various abnormal behavior modes of the engine and its components, which a priori have been known to exist and for which a model of the indications of abnormal behavior can be obtained from prior knowledge or training data obtained during such system behavior. Recognition of those indications leads to natural identification of the underlying condition and an appropriate action to mitigate that abnormal condition. In a way, this approach to realizing the condition monitoring functionality in an automotive engine essentially represents part of the engine design process, where the condition monitoring (just like the engine itself) is carefully designed through selection of hardware and algorithms that facilitate maximal diagnostic coverage of the abnormal behavior modes that are a priori known to exist. In the case of highly complex and dynamic systems, such as automotive or aircraft engines, the traditional approach to realizing diagnostic functionality becomes excessively cumbersome because of the need to train the condition monitoring processes to recognize a large number of faults, some of which often cannot be even anticipated in advance. Even for the cases one is able to anticipate in advance, many faults manifest themselves very differently under different driving inputs and environmental conditions, which makes training of diagnostic units for all possible conditions and all possible faults practically impossible.

A new paradigm inspired by the functionalities and characteristics of natural immune systems will be described in this paper. It is aimed at enabling continuous identification of abnormal system behavior, isolation of the source of anomalous behavior and performance recovery in spite of the presence of the anomaly in an automotive system. Realization of these functionalities should be pursued regardless of the driver inputs and external conditions, and without the need to *a priori* know the entire set of faults that the monitoring system is supposed to capture and mitigate. Fig. 1 illustrates analogy of the newly proposed paradigm with the natural immune systems.

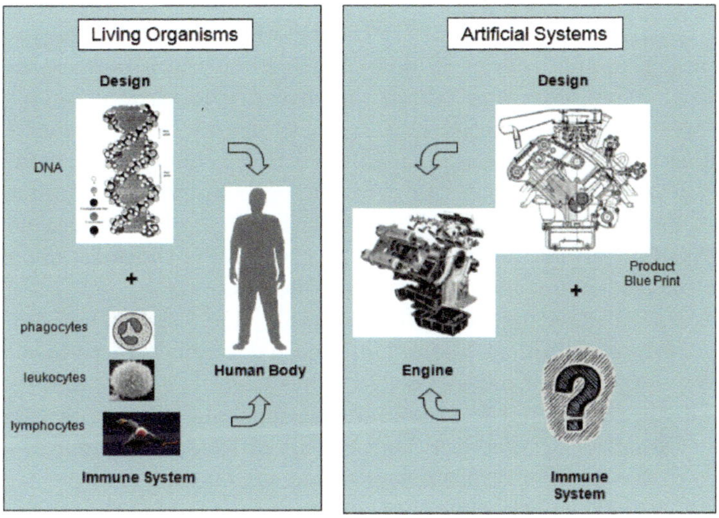

Fig. 1 Lack of immune systems in engines

This chapter outlines results of the on-going research to facilitate immunity-like reactions in automotive engines. These reactions will enable the vehicle to deal with faults that have not necessarily been observed in the past, using mechanisms similar to those used by a human body to deal with newly encountered viruses. The remainder of this chapter is organized as follows. Sect. 2 outlines research issues in immune systems engineering, denoting the proposed mechanism of immune system operation in an engine, methods for anomaly detection, fault localization and diagnosis. Sect. 3 describes an example of anomaly detection, fault localization and diagnosis demonstrated in high-fidelity simulations of an automotive electronic throttle system. These results will include an example of "fever-like" symptoms in an engine that encountered an unknown fault (an unknown "virus"). Sect. 4 describes an example of anomaly detection in crankshaft dynamics using a novel method of dynamic systems modeling. Finally, Sect. 5 offers conclusions of this work and outlines directions of future research.

2 Research Issues in Immune Systems Engineering

Anomaly detection can be formally described as identification of anomalous system behavior. Anomaly detection is one of the most important functions of a natural immune system. Before taking any actions to eliminate the foreign invaders, the immune system first has to discriminate self and harmful non-self. Natural immune systems are able to detect harmful non-self even when the virus or bacteria have never been encountered. This capability of being able to identify harmful non-self is vital for the survival of living organisms since the types of infections are very diverse and unpredictable.

A natural immune system is also able to learn the protein structures of pathogens it encounters and retain memory of the structures so that the future response to the same pathogens is faster and more efficient (4). This self-learning capability is important for natural immune systems to quickly adapt to a new environment. As a result, each individual immune system is unique. Furthermore, the memorized pathogen structures enable natural immune systems to correctly identify the type of pathogens (analogous to fault identification) and later take appropriate actions.

Once the pathogens are detected and possibly identified, a natural immune system will have to eliminate them. Different pathogens have to be eliminated in different ways (4). For pathogens that have never been encountered before, the initial response of a natural immune system is mild and it may take several weeks to eliminate the infection. During this process, the immune system is able to learn the protein structures of the newly encountered pathogens. On the other hand, if the pathogens or other similar types of pathogens have been encountered before, the responses of the natural immune system is usually so efficient that there are no clinical indications of a re-infection (4).

The above mentioned three functions (anomaly detection and localization, fault identification and self-healing) are essential for the survival of the living organism. The functionalities of the immune system are accomplished with the following desired characteristics: autonomy, robustness and uniqueness (4; 10). Autonomy of immune systems in living creatures displays itself in the ability to monitor and protect the living organism without outside control. The same mechanisms that monitor and protect the rest of the body also monitor and protect the immune system itself. Furthermore, the highly decentralized and distributed nature of an immune system and its ability to dynamically adapt to new situations constitute its robustness. Finally, the very adaptations that give an immune system its robustness provide each individual immune system unique characteristics. Since the immune system adapts to the intrusions, in spite of inherent similarities between individuals of one same species (individual machines of the same design), living (operating) conditions of each individual dictate the character of the immune system that the individual carries.

As mentioned earlier, an immune system paradigm is not currently implemented in automotive engines. Instead, automotive diagnostics is based on a predetermined set of comprehensive tests designed to capture specific, known faults. These tests are based on the prior knowledge about the system and there is no ability to capture new knowledge or to identify situations that were not covered by the diagnostic tests. In addition, the diagnostic tests serve solely the purpose of condition diagnosis, and their results are not used to attempt to eliminate the influence of a fault and restore the desired performance of the degraded automotive system.

Inspired by the analogy between the natural immune system of living organisms and diagnostic systems (in a more general sense) of vehicles, we summarize the functionalities of an automotive immune system in Fig. 2, suggesting that the immune system functionalities are realized through the functions of anomaly detection, fault localization, fault diagnosis and performance recovery.

An anomaly ("intrusion") in system performance is first identified as a statistically significant departure of system behavior away from that described by the model of normal system behavior. Subsequently, if an anomaly is found, multiple anomaly detectors connecting to relevant subsystems of the anomalous system are generated, with each detector of the anomalous behavior splitting even further into detectors monitoring subsystems with increased granularity. Such multiplication of anomaly detectors ultimately leads to localization of anomalous subsystems within the system, even if the underlying fault was not observed before, as shown in (11). Generation of additional anomaly detectors as an anomaly is detected mimics the way leucocytes are multiplied and flock around the intruders when an intrusion or illness is detected in a living organism.

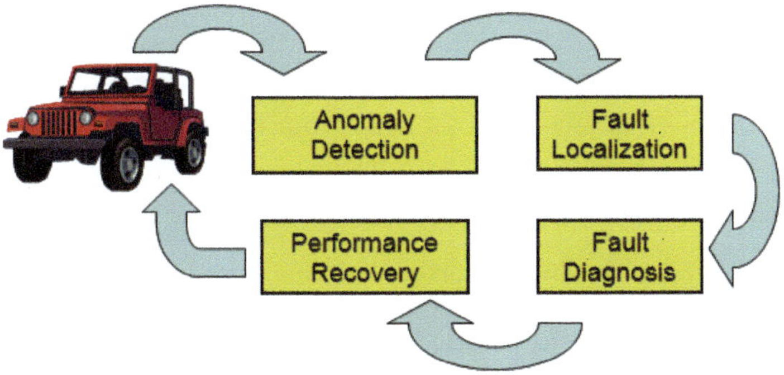

Fig. 2 Functionalities of an immune system in an automotive engine

Anomaly detection and localization are followed by fault diagnosis accomplished by matching the input-output patterns of the anomalous subsystem with the models of various previously seen faults. Each time the observed patterns could not be matched with any of the existing models, a new model needs to be created in order to recognize that particular situation in the future. This function is analogous to the process of generation of protein structures that match those of the intruding virus, after which antibodies with those protein structures can be synthesized to neutralize the virus.

Finally, based on the dynamic models identified through the diagnostic process, a control mechanism can be created to augment the nominal controllers in the system and restore as much as possible the original system function in the presence of a fault. This step resembles the generation of antibodies whose protein structures neutralize the intruding virus, thus effectively healing the organism.

Recently, it was suggested that the aforementioned immune system functionalities can be realized through anomaly detection agents, diagnostic agents and control agents. Notice that the separation of the three types of agents is only *conceptual* and does not have to be necessarily the case in a real implementation.

1. The role of Anomaly Detection Agents (ADA-s) will be to:
   - Detect and quantify any performance degradation that is manifested through input/output relationships
   - Localize the anomaly source to a specific subsystem or component by reconfiguring and reconnecting the anomaly detectors until input/output sets with anomalous behavior are identified in as much detail as possible.
2. The role of Diagnostic Agents (DA-s) will be to:
   - When the input-output patterns from the anomalous subsystem have been observed in the past and its input-output signatures have been memorized by a DA, correctly identify the type of anomaly.
   - If the input-output patterns characterizing the anomaly have never been encountered before, create a new DA that will be able to recognize and discriminate the observed anomaly from the set of known faults next time it occurs.
3. The role of Control Agents (CA-s) will be to:
   - Utilize fault characterization from DA-s in order to postulate control laws in such a way that desired performance of the anomalous subsystem is restored as much as possible. Hence, CA-s would effectively yield the self-healing functionality to the automotive system.

In the next three subsections, research issues and challenges associated with realization of the new diagnostic paradigm will be discussed in detail.

2.1 Anomaly Detection and Fault Localization

While conventional approaches to diagnosis focus on detecting and recognizing situations or behaviors which have previously been known to occur or can be anticipated, anomaly detection focuses on detecting and quantifying deviations away from learned "normal" behavior. The idea of anomaly detection has been successfully employed for building effective Intrusion Detection System (IDS) (5; 12; 13; 14), in which anomaly detection is used to identify malicious unknown attacks on a computer security system. Employing a similar idea in an automotive engine is not a straightforward task.

Identifying harmful non-self and quantifying the dissimilarity require an accurate definition of normal behavior. Therefore, constructing a profile representing normal behavior is essential for the success of the anomaly detection system. However, high complexity of the process dynamics of certain processes, such as the combustion process in an internal combustion engine, often prevents one from building an accurate model based on first principle. On the other hand, the process related signals are rather easy to obtain from sensors and embedded controllers. Consequently, data-driven approaches have been extensively employed in practice for modeling of complex dynamic systems. Among those techniques, neural networks, such as Multi-Layer Perceptron (MLP) networks (15), Radial Basis Function (RBF) networks (16) and Recurrent Neural Networks (RNNs) (17; 18; 19), are the most extensively applied techniques because of their universal function-approximating capabilities (15), particularly in the case of complex nonlinear systems. Unlike feed-forward networks such as RBF and MLP, which cannot identify temporal relationships in time series data without explicit time-delays introduced into each input vector, RNNs take into account temporal dependencies through local or global internal feedback connections in the network, which enables a good approximation of a wide class of nonlinear dynamical systems (15). However, the commonly used gradient descent algorithms to train RNNs exhibit problems during training, such as having difficulty dealing with long-term temporal dependencies in a time series (17), as well as "recency" difficulties (learning best what they have seen most recently). In addition, finding a suitable number of hidden neurons and an appropriate RNN structure remains a challenging problem. Since the training data for neural networks are usually taken from the entire operation space, they can be viewed as a global dynamic modeling approach (20). An alternative for modeling nonlinear dynamic system is the "divide and conquer" approach (21), which is based on the idea of dividing the whole system operation space into small sub-regions and modeling the local dynamics individually within each sub-region. The heuristic is that the modeling task for a small sub-region of the system behavior becomes easier to deal with than modeling the system as a whole. By dividing the operation space into small regions, the multiple local model approach can provide additional transparency about the physical processes, which is usually not easy to obtain for most of the global modeling approaches. This additional transparency is now also thought to be beneficial for later controller reconfiguration. Once an accurate model is obtained, the discrepancies between outputs of the model and the actual system can be utilized to detect and quantify the deviations. Since performance deviations are often caused by gradual component wear which may undergo a long developing process, quantification of anomaly severity is highly desirable for a number of applications, such as prognostics and health management systems (PHM) in aerospace industries (22; 23), rotating machinery remaining life estimation (2; 3) and condition-based maintenance (CBM) (24).

However, the interpretation of the residual errors between the outputs of the model and the actual system is still a challenging task. It is quite common in practice that the learned model is not perfect and there are always regions that the model cannot approximate well. Thus, the magnitudes of the output errors may vary when visiting different operational regions. As a result, surges in residual errors could merely be the results of switching from one operational region to another and do not necessarily indicate anomalies. Therefore, the decision making scheme has to be able to deal with these modeling uncertainties.

Localization of anomalies can be accomplished through distributing ADA-s throughout a complex dynamic system as shown in Fig. 3. Each ADA is only responsible for monitoring one particular subsystem and essentially decouples the subsystem from the rest of the complex dynamic system. Then, any significant deviation detected by an ADA only indicates that something abnormal occurs in the subsystem on which it is applied.

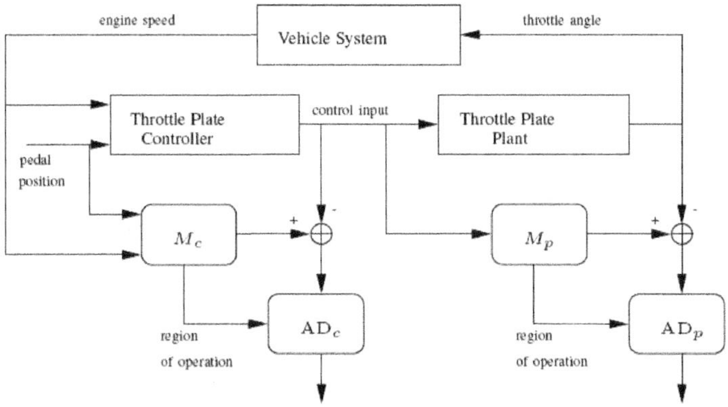

Fig. 3 Electronic throttle plate. M_c and M_p and are the model for the controller and plant respectively. AD_c and AD_p are the anomaly detectors for the controller and plant, respectively

2.2 Fault Diagnosis

Once the anomaly is detected and localized to a specific subsystem or component by the ADA-s, the next step is to identify the root causes of the anomalous system behavior.

This task can be accomplished using diagnostic agents that are connected to the inputs and outputs of the subsystem and recognize the anomalous condition as a particular fault if that fault has been observed before, or identify the observed situation as something that the immune system has not seen before. If a DA identifying the anomalous behavior pattern cannot be found, a new DA will be created for that new abnormal behavior mode.

The goal of each newly created DA will be not only to detect the presence of a particular faulty behavior mode, but also to parameterize that fault, by identifying the model of the new, degraded system if possible. This task is important in order to facilitate the creation of control agents that will augment the control system based on the newly identified faulty system model and compensate the effects of the degradation.

DA-s are envisioned in a manner similar to ADA-s. While ADA-s model normal behavior of the system and detect anomalies through identifying the departure of performance signatures away from those characterizing normal system behavior, DA-s will model a particular faulty behavior and can identify the occurrence of that fault in the system based on the proximity of signatures extracted out of the anomalous system to those characterizing that particular fault. In short, while ADA-s are envisioned as agents modeling normal system behavior and detecting departures away from it, DA-s are agents modeling a particular faulty behavior mode and detecting the proximity of current system behavior to that particular behavior mode. Challenges involved in the characterization of normal behavior of an automotive system, which were described in Sect. 2.1, very much apply to the case of modeling of any particular faulty behavior.

2.3 Automatic Control System Reconfiguration

Although the well designed feedback controller can typically reduce the plant's output sensitivity to measurement noise and disturbances, the degrees of failure that can be accommodated by a fixed control structure are restricted (24) and reconfiguration of the controller structure or its parameters based on the observed system faults or anomalies is usually necessary to maintain the required system performance.

Various approaches have been developed in the last two decades for fault accommodation and fault tolerant control (25), such as pseudo-inverse method (26), integral sliding mode control for sensor and actuator failures (27), linear quadratic regulator (LQR) (28) and autonomous control reconfiguration with parameter estimation (29). Despite the progress made up to date on fault accommodation, it was pointed out in (30) that due to the inherent complexity of nonlinear systems, most fault compensation methods rely primarily upon the well-developed linear control design methodology to achieve the required performance. This in turn limits their applicability to a number of real practical problems involving complex, nonlinear systems. For anticipated failures or faults, the control reconfiguration can be based on the stored control laws as proposed in (30; 31; 32). The real challenge resides in the full automation of the process of generation of new controllers for general nonlinear systems when an unanticipated fault occurs.

As described in Sect. 2.1 and Sect. 2.2, the occurrence of an unknown fault that cannot be recognized by any of the existing DA-s should cause creation of a new DA in the automotive immune system, which will model the newly observed behavior mode. This model will then be utilized to create a CA in the shape of a non-linear controller that will compensate for the negative effects of the observed anomaly.

3 Anomaly Detection, Fault Isolation and Diagnosis in an Automotive Electronic Throttle System

An example of anomaly detection, fault localization and fault diagnosis in an automotive electronic throttle control system will be provided in this section. A high fidelity software package provided by a major manufacturer of software for automotive control system development is used in the experiments. Fig. 3 shows the block diagram of the simulation software. Note that the controller used in the simulation can be readily embedded into microprocessors. We introduce the use of two anomaly detectors in a distributed fashion as illustrated in Fig. 3. These two anomaly detectors essentially decouple the plant and the controller.

The signals are collected at the sampling rate of 100Hz. The anomalies are simulated by modifying the parameters in the controller and the plant. Different levels of parameter changes are introduced to simulate different levels of degradation. Simulations are run using various standard driving profiles. To investigate the effects of noise, 5% multiplicative noise has been imposed on the outputs of the controller and the plant. This corresponds to a signal to noise ratio of 26dB.

3.1 Anomaly Detection and Fault Isolation

To demonstrate the capability of fault isolation through distributed anomaly detection without the need to *a priori* know signatures characterizing a given fault, two anomaly detectors are applied separately on the controller and the plant as shown in Fig. 3. Three different test scenarios are introduced by gradually changing the actual system parameters over time to simulate the system degradation, as illustrated in Fig. 4. In each scenario only one parameter is varied in time while all the other parameters are kept at their nominal value.

Each anomaly detector was configured to output overlaps between modeling distribution errors associated with normal system behavior and those associated with the most recent system behavior, as illustrated in Fig. 5 Following (33), these overlaps are referred to as the performance Confidence Values (CVs). The usefulness of the concept of CVs is in the fact that high CVs (close to 1) indicate performance close to nominal, while low levels of CVs (close to 0) indicate degraded system performance.

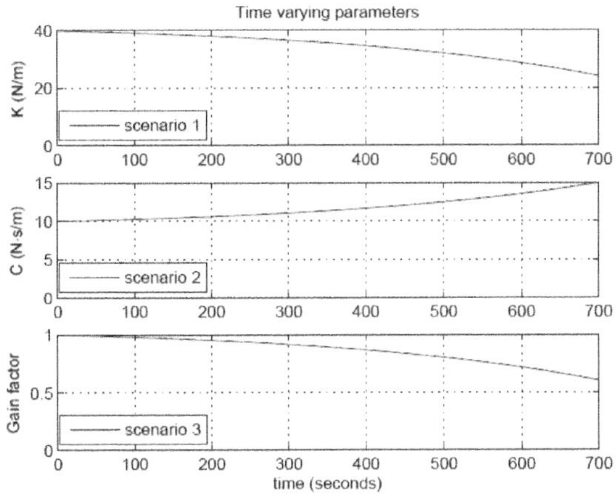

Fig. 4 Time varying parameters for scenarios 1, 2 and 3. The nominal values for stiffness (plant parameter), viscous damping (plant parameter) and gain (controller parameter) are 40N/m, 10Ns/m and 1, respectively

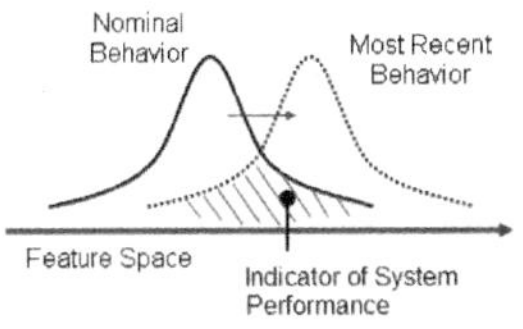

Fig. 5 Performance Confidence Value (CV) as a system performance indicator based on the overlap of signature distributions corresponding to nominal and the most recent system behavior

Fig. 6 shows the outputs of the two anomaly detector for the three introduced scenarios. The straight line across the window is the steady state lower 3σ control limits of the EWMA control chart applied to the performance CVs obtained during training. It can be easily observed from Fig. 5 that in Scenario 1, the CVs from the anomaly detector operating on the controller are high all the time, while the CVs from the anomaly detector on the plant gradually decrease and finally exceed the control limit. This indicates that an anomaly has occurred in the plant, while the controller is still operating normally. Similar behavior can also be observed in Scenario 2. In Scenario 3, we have introduced a disturbance only to the controller and, as expected, the performance CVs from the controller anomaly detector eventually exceed the control limits, whereas the performance CVs from the plant detector remain within the control limits. In this manner, we can readily localize the anomalous behavior to the plant, the controller or both.

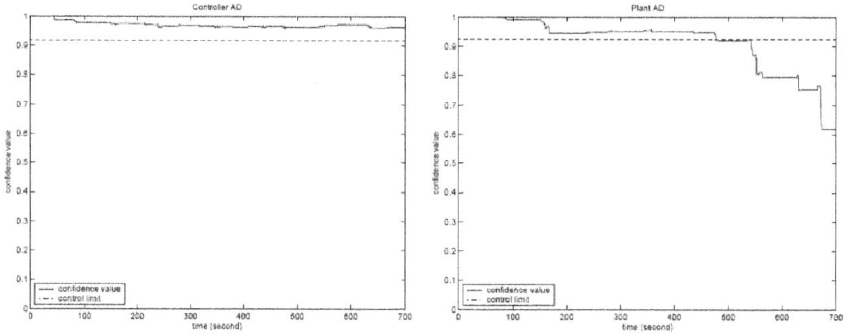

(a) Scenario 1: Stiffness K (plant parameter) decreases from its normal value 40N/m to 24N/m in 700s.

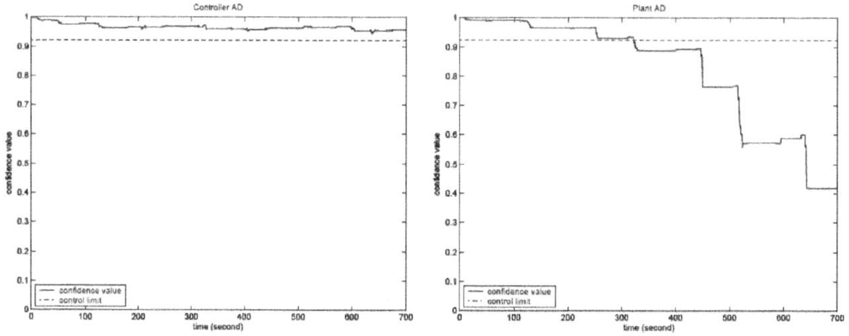

(b) Scenario 2: Viscous damping C (plant parameter) increases from its normal value 10N·s/m to 15N·s/m in 700s.

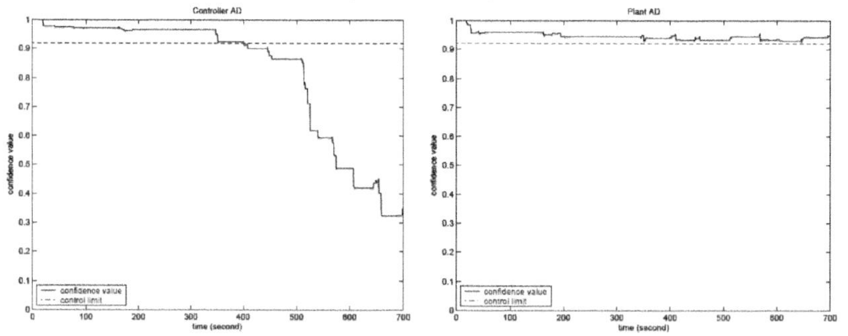

(c) Scenario 3: Controller gain decreases from its normal value of 1 to 0.6 in 700s.

Fig. 6 Experimental results for the three designed scenarios. Note that the confidence values are updated only when the anomaly detectors are triggered

3.2 Fault Diagnosis

The root cause identification can be accomplished by several means. The simplest way of fault isolation is through the use of multiple anomaly detectors distributed throughout a control system. The anomaly can be localized through the identification of the lowest level segment of the system in which an anomaly detector has announced the presence of anomalous behavior. This scheme is used in the previous section to isolate the faults between the controller and the plant. A straightforward extension of this approach creates a second method of root cause identification in which specialized diagnosers identify specific failure modes. In this approach, separate diagnosers are specifically trained to detect specific faults. The fault is then isolated by the simultaneous drop in the confidence level of the anomaly detector (measuring proximity of system behavior to the normal behavior), along with the growth in the confidence level of the diagnoser associated with the actual fault (indicating proximity of system behavior to behavior corresponding to that particular fault). This approach is illustrated in the following example. Table 1 shows the operating conditions for various normal and failure modes that were simulated. It was assumed that the input-output signals from only the states F0, F1 and F2 are available for training, while signatures from the state F3 were unknown. Three detectors corresponding to states F0, F1 and F2 were trained as described above, using data obtained from simulations of standard US and European test driving profiles. Each of the detectors is then used to describe the statistical profiles of a particular plant condition.

Table 1 Labels for various simulated situations

Fault Label	Description
F0	Normal
F1	Reduced K
F2	Increased C
F3	Unknown fault (reduced saturation limit on throttle motion, which corresponds to dirt accumulation in the throttle chamber

Fig. 7(a) shows the time intervals during which operating states F0, F1, F2 and F3 were introduced. Fig. 7(b) shows the confidence values calculated using the diagnosers of the three known situations. In each case, the appropriate diagnoser registers the occurrence of the correct condition within the proper interval. When the unknown condition F3 appears, the low CVs from all diagnosers indicate the presence of an anomaly that has not been seen before.

3.3 Fever-Like Behavior in the Presence of an Unknown Fault

In this subsection, we will demonstrate results obtained by tracking the computational effort (floating point operations – FLOPS) when an unknown fault was introduced into the electronic throttle system. Initially, all system

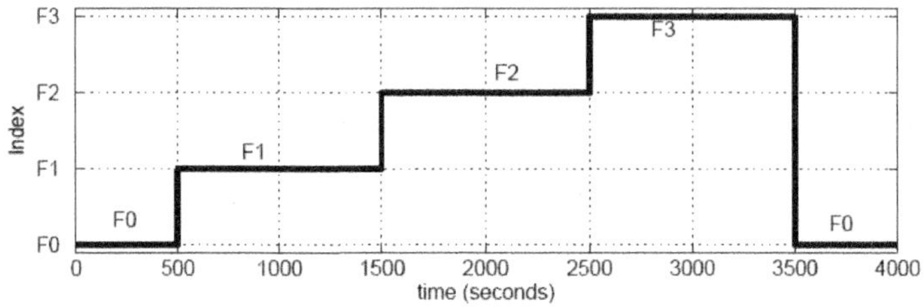

(a) Sequence of simulated states in the electronic throttle system.

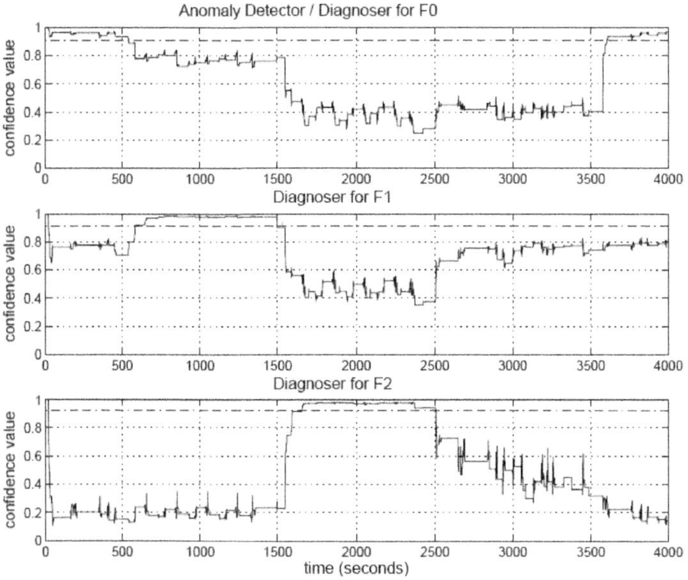

(b) Performance confidence values calculated from diagnosers specialized for states F0, F1, and F2.

Fig. 7 Fault diagnosis in a simulated automotive electronic throttle system

parameters were set to nominal values and a global anomaly detector was monitoring the system performance. At 50 seconds, a 20% random perturbation in plant and controller parameters was introduced, which caused the global anomaly detector to detect an anomaly and split into 2 anomaly detectors – one monitoring the controller and one monitoring the pant in the electronic throttle system (as illustrated in Fig. 8). Subsequently, both anomaly detectors detected significant departure of modeling errors away from normality, thus detecting the presence of anomalies in both the plant and the controller. This initiates creation of dynamic models corresponding to the anomalous subsystems. All models were created using the recently introduced Growing Structure Multiple Model System (GSMMS) method (34), which will be briefly described in the next section. Over time, both models converge and the system now has "memory" of this specific anomalous behavior. During this process of generation of anomaly detectors and GSMMS models of anomalous behavior, one can observe a surge in computational effort, which subsided only when the model of the fault was established, as illustrated in Fig. 9. This surge in computational effort triggered by detection of an unknown condition in the system ("organism") can be viewed as a fever occurring in the automotive engines and its ultimate product – a model of a newly encountered fault – can be seen as a portion of the ultimate antibody that will cure the engine from this fault (appropriate controller for recovering the performance being the second part of this antibody). Creation of full antibodies for immunity in engines remains a challenge for future research.

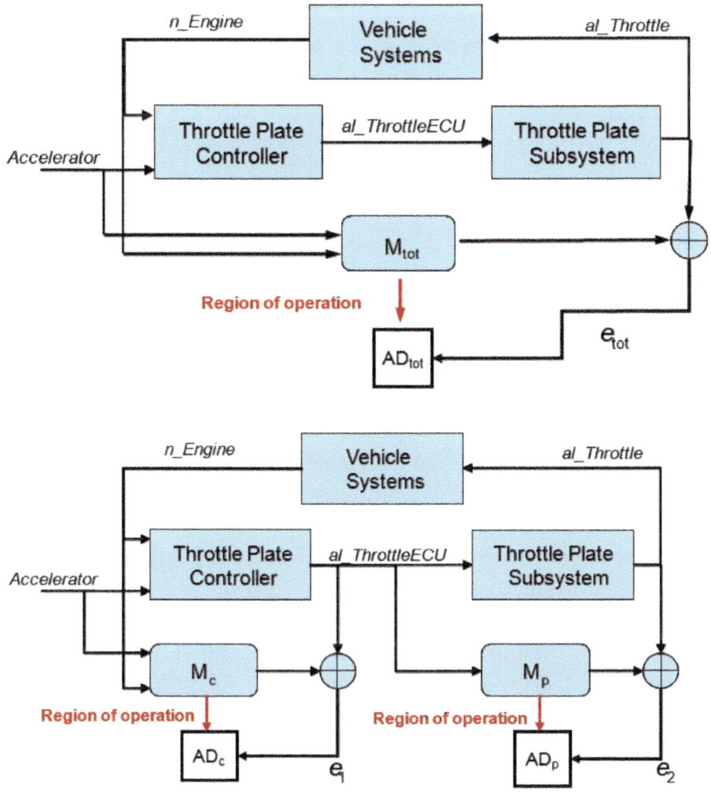

Fig. 8 Splitting of a global anomaly detector (AD_{tot}) into anomaly detectors for the controller and the plant (AD_c and AD_p, respectively) when an unknown fault is introduced into the electronic throttle system. M_{tot}, M_c, and M_p denote GSMMS models of the entire system, controller and the plant, respectively. Labels e_{tot}, e_1 and e_2 denote GSMMS modeling errors emanating from models of the entire system, controller and the plant, respectively

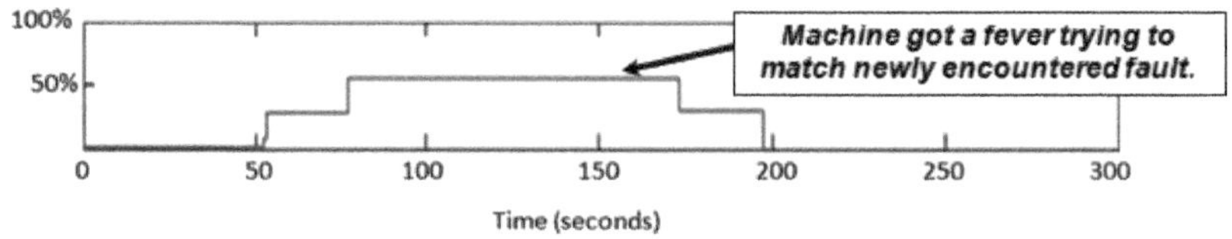

Fig. 9 Surge in relative computational effort due to an unknown fault introduced into the plant and controller of an automotive electronic throttle controller. Surge in the computational effort is ended when the GSMMS models of the faulty subsystems converged

4 Anomaly Detection and Fault Isolation in Automotive Crankshaft Dynamics

Another application example uses Growing Structure Multiple Model System (GSMMS) based anomaly detection to monitor the crankshaft dynamics. GSMMS combines the advantages of the growing Self-Organizing Networks (SONs) (35; 36) with efficient linear model parameter estimation algorithms (37) into an integrated framework for modeling and identification of general nonlinear dynamic systems. GSMMS can be implemented on-line to learn the structural parameters and the local model parameters simultaneously. Generally speaking, one starts from relatively small number of nodes (or equivalently local models) and then gradually adds nodes to the ordered network where the fitting error is high. The training procedure will terminate once a stopping criterion, such as desired modeling residuals level, or maximal number of SOM nodes, is satisfied. This way, the number of nodes can be determined automatically, without guesswork. For more detailed information about the GSMMS, please refer to (34), and references therein. A set of training data is first collected from the simulation of the first principle model of crankshaft dynamics during normal operation. From each combustion cycle corresponding to two revolutions of the crankshaft, eight samples are collected. This is equivalent to the sampling period of $\pi/2$ in the crankshaft angle domain. Based on a batch training algorithm proposed in (34), a GSMMS model is created. Fig. 10(a) compares the outputs of the first principle model and the trained GSMMS model when the viscous damping coefficient c_f is set to its nominal value. The GSMMS model can roughly capture the transient behavior of the crankshaft dynamics despite the existence of multiplicative noise. Fig. 10(b) also shows the resulting residual errors when c_f is increased by 20%. In comparison with Fig. 10(a), the increases in magnitudes of the residual errors are clearly visible at low engine speed. Once the GSMMS model is obtained from normal training data, the residual errors can then be utilized to detect and quantify performance deviations or anomalies as illustrated in Fig. 10. However, as we mentioned earlier, the interpretation of the residual errors between the outputs of the model and real system is not a trivial task since the magnitudes of the modeling errors may vary at different

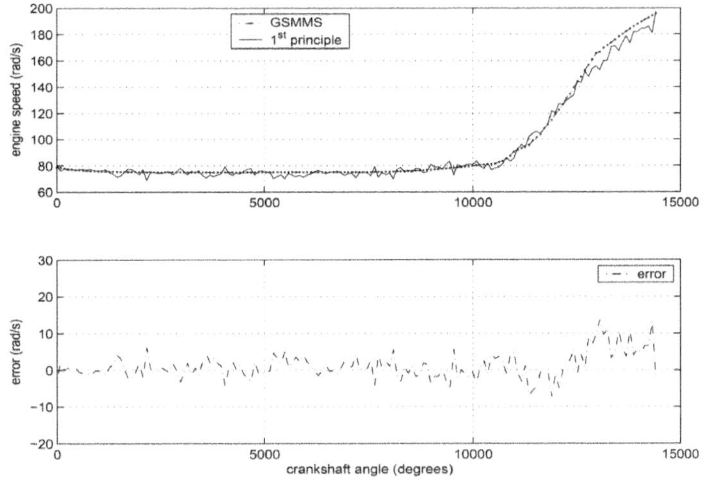

(a) $c_f = 0.0175\text{Ns/mm}$

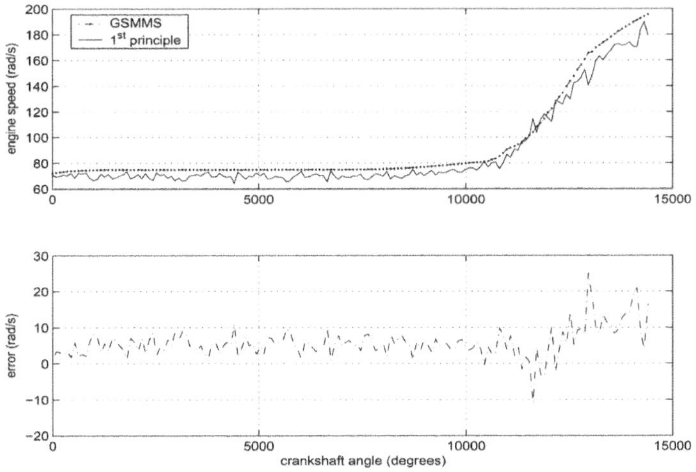

(b) $c_f = 0.021\text{Ns/mm}$

Fig. 10 Comparison of the outputs of the model obtained using first principles (used to train the GSMMS model) and the trained GSMMS model

operational regions. We can observe this from Fig. 10(a), where at low engine speed the magnitudes of the residual errors are substantially smaller than those at high engine speed. As a result, surges in residual errors could merely be the results of shifting from one operational region to another and not necessarily indicate anomalies. A region dependent decision making scheme is proposed in (34; 38) to overcome this problem and to improve the detection accuracy. The schematic overview of the newly proposed approach is shown

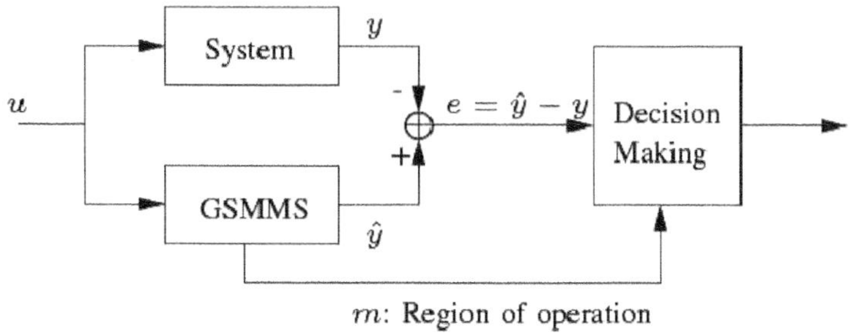

Fig. 11 Illustration of the region dependent decision making scheme

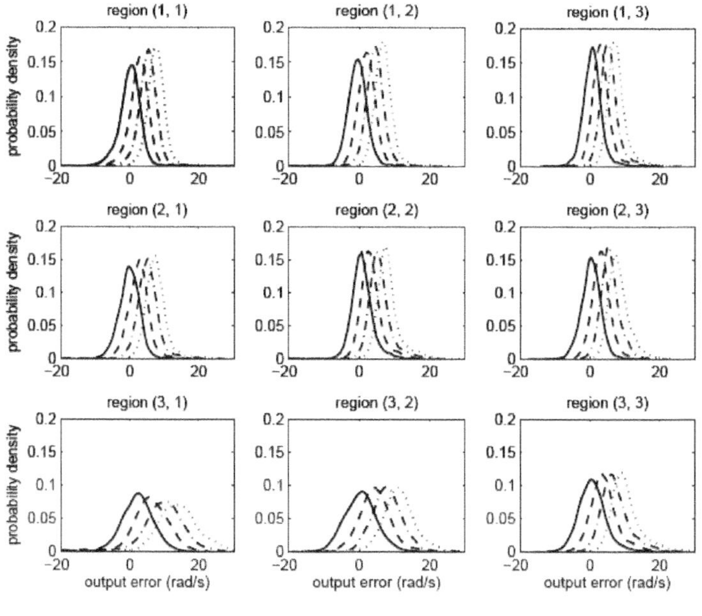

Fig. 12 Probability densities of output errors in selected nine adjacent regions. Solid line: $c_f = 0.0175$Ns/mm, which is the nominal value; Dashed line: $c_f = 0.01925$Ns/mm; Dash-dotted line: $c_f = 0.021$Ns/mm; Dotted line: $c_f = 0.02275$Ns/mm

in Fig. 11, where the residual errors are analyzed separately in each operating region (each Voronoi set of the SON underlying the GSMMS model). Fig. 12 shows the error distributions in different regions under different values of viscous damping coefficient c_f. We can observe that as c_f shifts away from its nominal value $c_f = 0.0175$Ns/mm, the error distributions also shift away from the corresponding distributions that represent normal behavior.

If only the detection of anomalies is of interest, Exponentially Weighted Moving Average (EWMA) control charts (39) can be employed in each operational region to perform this task by detecting out-of-limit residual errors. By varying the exponential forgetting factor, we can also investigate the performance of the anomaly detection systems for a given mean time to detection[1], since the mean time to detection can be determined by the values of the forgetting factor used in constructing EWMA control chart (39). Essentially, the EWMA control chart in each region forms an anomaly detector, which is responsible for detecting anomalous behavior once the system operates in that region. Receiver Operating Characteristic (ROC) curves (40) are utilized in this paper to evaluate the performance of anomaly detection. Fig. 13 shows the ROC curves for detecting a 10% increase in the viscous damping coefficients c_f given different mean time to detection. The nine regions shown in Fig. 13 are the same as those shown in Fig. 12. We can observe that the performance of the anomaly detectors in all nine regions keeps improving as more samples are utilized. Almost perfect detection can be achieved from all

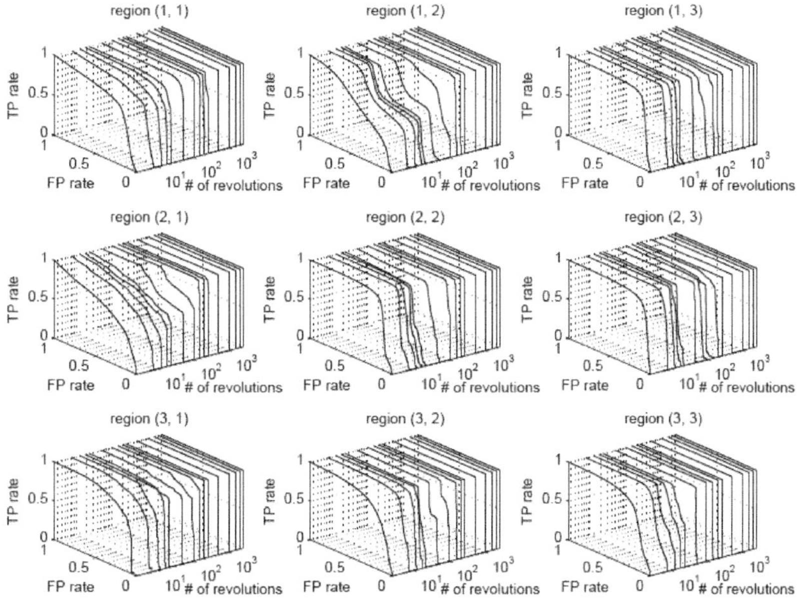

Fig. 13 Receiver Operating Characteristic (ROC) curves from selected nine adjacent SON regions, which are the same as those shown in Fig. 10. The ROC curves are plotted against the amount of data used (8 samples per revolution). TP rate denotes the true positive rate in classification, while FP rate denotes the false positive rate in classification

[1] Since the crankshaft dynamics are expressed in the crankshaft angle domain, the mean time to detection here is referred to as the number of rotations.

different regions when data length reaches about 1000 revolutions. Similarly observations have also been made for detecting 20% and 30% increases in c_f. As expected, the mean time to detection is substantially shorter for achieving a similar detection accuracy.

We have observed that as c_f shifts away from its nominal value the error distributions also shift away from the distributions that represent normal behavior. As a result, the overlaps become smaller as the severity of the anomaly increases. Therefore, one way to quantify the severity of the anomaly is to evaluate the overlaps between current error distributions and the error distributions under normal operation, as was done in (34).

5 Conclusions and Future Work

In this chapter, we propose a novel method for anomaly detection, fault localization and diagnosis, inspired by the way these functions are accomplished in natural systems. Anomalies are detected as statistically departures away from normal behavioral patterns and are localized to the culprit system(s) through distributed anomaly detection that delves into subsystems of deeper granularity each time an anomaly is detected. Fault diagnosis is accomplished through the traditional diagnostic approaches of matching model indications of known faults with signatures of the currently observed system behavior.

The key enabling method facilitating the above-mentioned vision are the newly developed generic approaches to modeling of complex system dynamics (11; 34), allowing anomaly detectors and diagnosers to use essentially the same modeling mechanism to successively connect to different inputs and outputs (corresponding to different subsystems), identify models of the corresponding subsystems and accomplish anomaly detection and diagnosis. The new dynamic modeling approaches are based on representing complex system dynamics using the "divide and conquer" philosophy. Description of system dynamics is cast into the framework of connected multiple models, each of which is relatively simple in nature and can be analyzed in an analytically tractable manner. In (11), local statistical models of moments extracted from time-frequency distributions of output signals in each SON Voronoi set are used, while in (34) we postulated local linear dynamic models. Examples of anomaly detection, fault localization and diagnosis in automotive electronic throttle system and crankshaft dynamics are presented. Previously unknown faults are localized to appropriate subsystems and "fever-like" symptoms of surging computational effort are observed as the "automotive immune system" struggled to characterize the newly encountered fault. All results are obtained using high fidelity automotive system simulations developed by a major manufacturer of software for automotive control and diagnostic applications.

Future research will be dedicated to utilizing local model tractability emanating from the "divide and conquer" modeling approaches to devise methods for controller adaptation that will facilitate performance recovery in the presence of faults. Ability to adaptively detect, localize, characterize and compensate for faults will enable one to explore possibilities of artificially vaccinating vehicles against different faults by inserting those faults into a selected group of "test-vehicles" that will develop immunity to those faults and whose "antibodies" could then be distributed into the entire fleet of vehicles as diagnostic and control software patches. We also believe that immune systems inspiration can have far fetching influence in future developments in other engineering areas, such as manufacturing, aerospace and biomedical systems.

Acknowledgement

This work was supported in part by ETAS Inc. and National Science Foundation (NSF) under Grant CMMI 0829237. Any opinions, findings and conclusions or recommendations expressed in this material are those of the authors and do not necessarily reflect the views of the NSF.

References

1. Rao, B.: Handbook of Condition Monitoring. Elsevier Advanced Technology (1996)
2. Randall, R.B.: State of the art in monitoring rotating machinery – part 1. Sound and Vibration 38(3), 14–21 (2004)
3. Randall, R.B.: State of the art in monitoring rotating machinery - part 2. Sound and Vibration 38(5), 10–17 (2004)
4. Hofmeyr, S., Forrest, S.: Architecture for an artificial immune system. Evolutionary Computation 8(4), 443–473 (2000)
5. Denning, D.: An intrusion detection model. IEEE Transactions of Software Engineering, SE-13(2), 222–232 (1987)
6. Kim, Y.-W., Rizzoni, G., Utkin, V.: Automotive engine diagnosis and control via nonlinear estimation. IEEE Control Systems Magazine 18(5), 84–97 (1998)
7. Patton, R., Frank, P.M., Clark, R.N.: Issues of Fault Diagnosis for Dynamic Systems. Springer, New York (1996)
8. Venkatasubramanian, V., Rengaswamy, R., Yin, K., Kavuri, S.N.: A review of process fault detection and diagnosis part I: Quantitative model-based methods. Computers and Chemical Engineering 27(3), 293–311 (2003)
9. Venkatasubramanian, V., Rengaswamy, R., Kavuri, S.N.: A review of process fault detection and diagnosis part II: Qualitative models and search strategies. Computers and Chemical Engineering 27(3), 313–326 (2003)
10. Brotherton, T., Johnson, T.: Anomaly detection for advanced military aircraft using neural networks. In: Proc. of the IEEE Aerospace Conference, vol. 6, pp. 3113–3123 (2001)
11. Liu, J., Djurdjanovic, D., Marko, K.A., Ni, J.: A Novel Method for Anomaly Detection, Localization and Fault Isolation for Dynamic Control Systems. Submitted to Mechanical Systems and Signal Processing (in review), Paper no. MSSP07-153 (2007)

12. Anderson, J.P.: Computer security threat monitoring and surveillance. Technical Report, James P. Anderson Company (1980)
13. Estevez-Tapiador, J.M., Garcia-Teodoro, P., Diaz-Verdejo, J.E.: Anomaly detection methods in wired networks: A survey and taxonomy. Computer Communications 27(16), 1569–1584 (2004)
14. Lunt, T.F.: Survey of intrusion detection techniques. Computers & Security 12(4), 405–418 (1993)
15. Haykin, S.S.: Neural networks: a comprehensive foundation, 2nd edn. Prentice Hall, Upper Saddle River (1999)
16. Cho, K.B., Wang, B.H.: Radial basis function based adaptive fuzzy systems and their applications to system identification and prediction. Fuzzy Sets and Systems 83(3), 325–339 (1996)
17. Han, M., Xi, J., Xu, S., Yin, F.-L.: Prediction of chaotic time series based on the recurrent predictor neural network. IEEE Transactions on Signal Processing 52(12), 3409–3416 (2004)
18. Leung, H., Lo, T., Wang, S.: Prediction of noisy chaotic time series using an optimal radial basis function neural network. IEEE Transactions on Neural Networks 12(5), 1163–1172 (2001)
19. Connor, J.T., Martin, R.D., Atlas, L.: Recurrent neural networks and robust time series prediction. IEEE Transactions on Neural Networks 5(2), 240–253 (1994)
20. Principe, J.C., Wang, L., Motter, M.A.: Local dynamic modeling with self-organizing maps and applications to nonlinear system identification and control. Proc. of the IEEE Conference 86(11), 2240–2258 (1998)
21. Kim, S.-P., Sanchez, J.C., Erdogmus, D., Rao, Y.N., Wessberg, J., Principe, J.C., Nicolelis, M.: Divide-and-conquer approach for brain machine interfaces: Nonlinear mixture of competitive linear models. Neural Networks 16(5-6), 865–871 (2003)
22. Roemer, M., Nwadiogbu, E., Bloor, G.: Development of diagnostic and prognostic technologies for aerospace health management applications. In: Proc. of 2001 IEEE Aerospace Conference, vol. 6, pp. 3139–3147 (2001)
23. Brotherton, T., Grabill, P., Wroblewski, D., Friend, R., Sotomayer, B., Berry, J.: A testbed for data fusion for engine diagnostics and prognostics. In: Proc. of 2002 IEEE Aerospace Conference, vol. 6, pp. 3029–3042 (2002)
24. Hosseini, M.M., Kerr, R.M., Randall, R.B.: Inspection model with minimal and major maintenance for a system with deterioration and Poisson failures. IEEE Transactions on Reliability 49(1), 88–98 (2000)
25. Stengel, R.: Intelligent failure-tolerant control. In: Proc. of the 5th IEEE International Symposium on Intelligent Control, pp. 548–557. IEEE Comput. Soc. Press, Los Alamitos (1990)
26. Gao, Z., Antsaklis, P.: On the stability of the pseudo-inverse method for reconfigurable control systems. In: Proc. of the IEEE National Aerospace and Electronics Conference, pp. 333–337 (1989)
27. Kim, Y.-W., Rizzoni, G., Utkin, V.: Developing a fault tolerant powertrain control system by integrating design of control and diagnostics. International Journal of Robust and Nonlinear Control 11(11), 1095–1114 (2001)

28. Noura, H., Ponsart, J.-C., Theilliol, D.: Sensor fault-tolerant control method applied to a winding machine. In: Proc. of Fault Detection, Supervision and Safety for Technical Processes, vol. 3, pp. 777–782. Elsevier Sci., Amsterdam (2001)
29. Rauch, H.: Autonomous control reconfiguration. IEEE Control Systems Magazine 15(6), 37–48 (1995)
30. Yen, G., Ho, L.-W.: On-line intelligent fault-tolerant control for catastrophic system failures. In: Proc. of the SPIE, Component and Systems Diagnostics, Prognosis, and Health Management, vol. 4389, pp. 35–46 (2001)
31. Diallo, D., Benbouzid, M., Makouf, A.: A fault-tolerant control architecture for induction motor drives in automotive applications. IEEE Transactions on Vehicular Technology 53(6), 1847–1855 (2004)
32. Ichtev, A., Hellendoorn, J., Babuska, R., Mollov, S.: Fault-tolerant model-based predictive control using multiple takagi-sugeno fuzzy models. In: 2002 IEEE International Conference on Fuzzy Systems, vol. 1, pp. 346–351 (2002)
33. Lee, J.: Machine Performance Monitoring and Proactive Maintenance in Computer-Integrated Manufacturing: Review and Perspective. International Journal of Computer Integrated Manufacturing 8(5), 370–380 (1995)
34. Liu, J., Djurdjanovic, D., Marko, K.A., Ni, J.: Growing structure multiple model system for anomaly detection and fault diagnosis. To appear in Transactions of ASME, Journal of Dynamic Systems, Measurements and Control, Paper no. DS-06-1152 (2009)
35. Fritzke, B.: Growing cell structures - a self-organizing network for unsupervised and supervised learning. Neural Networks 7(9), 1441–1460 (1994)
36. Fritzke, B.: A growing neural gas network learns topologies. In: Proc. of Advances in Neural Information Processing Systems, vol. 7, pp. 625–632. MIT Press, Cambridge (1995)
37. Ikonen, E., Najim, K.: Advanced process identification and control. Marcel Dekker, New York (2002)
38. Liu, J., Sun, P., Djurdjanovic, D., Marko, K.A., Ni, J.: Growing structure multiple model system based anomaly detection for crankshaft monitoring. In: Proc. of International Symposium on Neural Networks (2006)
39. Montgomery, D.C.: Introduction to statistical quality control, 4th edn. John Wiley, New York (2001)
40. Duda, R.O., Hart, P.E., Stork, D.G.: Pattern Classification. Wiley, New York (2001)

Pitch-Depth Control of Submarine Operating in Shallow Water via Neuro-adaptive Approach

Y.D. Song, Liguo Weng, and Medorian D. Gheorghiu

Abstract. This chapter presented a neuro-adaptive control for depth and pitch control of submarine operating in shallow waters. The control scheme is based on two neural-network (NN) units which are shown to be effective in attenuating the reconstruction error and other lumped system uncertainties. Stable on-line weight-tuning algorithms are derived based on Lyapunov's stability theory. Variable heading speed and changing center of gravity are taken into consideration in control design. Theoretical analyses and simulations prove the efficacy of the proposed control scheme in dealing with external disturbances as well as system nonlinearities, uncertainties, and parameter variations.

1 Introduction

The primary objective of this research is to design robust and adaptive control algorithms for submarine operating in shallow waters (see Fig. 1). One important platform in littoral warfare is the submarine with the capabilities to move close to shore without revealing its presence; therefore it is an excellent vehicle for running reconnaissance missions, stealthily transporting SEAL for penetration operations, or launching a rapid, highly accurate missile strike (1), (7; 8; 9; 10). This work is concerned with the development of control schemes for submarines operating in shallow waters, able to track a desired pitch angle θ and a desired and safe depth z, thus reducing the possibilities of human error. Technically, operation of submarines is very complex due to multi-input multi-output nonlinear structure. The operations become even more complicated in the event of change in gravitational center, variation of heading speed or system failures, such as stern, bow plane jam, or actuator fault.

Because of its complexity, little research has been done regarding to the problem of submarine control (12; 13; 14). Furthermore, the present work normally either oversimplified the nonlinear model of submarines or ignored

J. Schumann, Y. Liu (Eds.): Appl. of Neural Networks in High Assur. Sys., SCI 268, pp. 165–178.
springerlink.com

Fig. 1 In shallow water (11)

the disturbances, therefore, to design a practical and effective controller for submarine maneuver remains an elusive goal.

In this chapter, we consider the nonlinear dynamic model coupled with unpredictable disturbances and uncertainties. We also take into consideration the shift in gravity center of the submarine. Because the mass of the crew operating the submarine can reach a significant percentage of the total mass of the vehicle, a change in the position of the center of gravity (***CG***) of the submarine is explicitly considered in control design. The shift of ***CG*** can occur in the event that part or all of the personnel is gathering in one side of the submarine, creating a moment arm with respect to the center of buoyancy (***CB***). With the buoyant force pushing up and the mass of the ship pushing down, a submersible tends to pitch.

2 Dynamics

For the problem under consideration, we focus on the motion (see Fig. 2) in vertical plane and treat the motions in other axis as disturbances.

Following Papoulias (2), we can derive Newton's equations of motion for heave and pitch as follows,

$$\begin{aligned} m[\dot{w} - uq - x_G\dot{q} - z_Gq^2] = Z_{\dot{q}}\dot{q} + Z_{\dot{w}}\dot{w} + Z_q uq + Z_w uw + \\ u^2(Z_b\delta_b + Z_s\delta_s) - C_D \int_{tail}^{nose} l_b(\tau)(w - \tau q)|w - \tau q| d\tau \end{aligned} \tag{1}$$

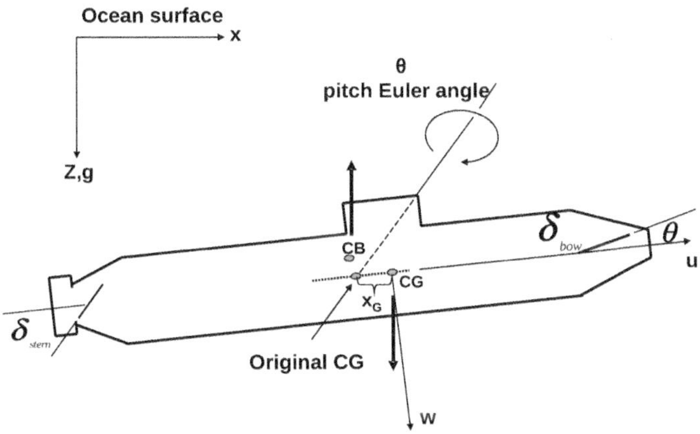

Fig. 2 Moment arm created by change in CG position

$$\begin{aligned} I_y\dot{q} - m[x_G(\dot{w} - uq) - Z_G(\dot{u} + wq)] = M_{\dot{q}}\dot{q} + M_{\dot{w}}\dot{w} + M_q uq + \\ M_w uw + u^2(M_b\delta_b + M_s\delta_s) - (x_G mg - x_B B)\cos(\theta) - \\ (z_G mg - z_B B)\sin(\theta) + C_D \int_{tail}^{nose} l_b(\tau)(w - \tau q)|w - \tau q|\tau d\tau \end{aligned} \tag{2}$$

where z is the depth, θ is the pitch angle, m denotes the vehicle mass including water contained in enclosed free-flood spaces, neutral ballast of the submarine, w is the heave speed, q is the pitch rate, u is the submarine velocity along longitudinal axis, δ_b and δ_s are the control surface deflections angles for the bow plane and stern plane respectively. Other coefficients can be found from (3).

Note that the terms involving the derivatives of w and q in the above equations are collected. The relation between θ and q is:

$$\dot{\theta} = q \tag{3}$$

and the motion in $z$direction is determined by:

$$\dot{z} = w\cos\theta - u\sin\theta \tag{4}$$

External forces and moments (i.e., wave forces, changes in trim) which act as disturbances against the origin of the submarine fixed coordinates and disturbances due to motion from other axis are also included in the model.

2.1 Nonlinear Model

For simple notation, we introduce $h = [z\ \theta]^T$ and re-express the model as follows:

$$\ddot{h} = E(z, \theta, u, \dot{z}, \dot{\theta}, \dot{u}) + Bv \tag{5}$$

The dynamic system equation described in (5) is controllable only if the matrix of B is invertible:

$$b_{11}b_{22}u^4\cos\theta - b_{12}b_{21}u^4\cos\theta \neq 0 \tag{6}$$

For a submarine with forward speed ($u \neq 0$) and non-vertical position ($\theta \neq 90°$), it can be shown that (6) is always true. In fact, the following Lemma can be stated:

Lemma: For any $u \neq 0$ and $\theta \neq \pm(2k+l)\pi/2$ the matrix B is invertible.

Proof: The determinant of B is:

$$\det(B) = u^4(b_{11}b_{22} - b_{12}b_{21})\cos\theta \tag{7}$$

It is seen that as long as $u \neq 0$, $\theta \neq \pm(2k+l)\pi/2$ and $u^4(b_{11}b_{22}-b_{12}b_{21})\cos\theta \neq 0$, then B is invertible. Substituting b_{11}, b_{12}, b_{21} and b_{22}, results in:

$$b_{11}b_{22} - b_{12}b_{21} = \frac{Z_bM_s - Z_sM_b}{(m - Z_{\dot{w}})(I_y - M_{\dot{q}}) - Z_{\dot{q}}M_{\dot{w}}} \tag{8}$$

For the submarine under consideration, the relations between forces and moments of the control planes are: $Z_b = Z_s/2$, $M_b = -M_s/4$. Namely, the force coefficients of the stern planes are doubled to provide a realistic level force, and bow planes have one half the force and one quarter the moment authority of the stern planes. Substituting into(8):

$$b_{11}b_{22} - b_{12}b_{21} = \frac{3}{4}\frac{Z_sM_s}{(m - Z_{\dot{w}})(I_y - M_{\dot{q}}) - Z_{\dot{q}}M_{\dot{w}}} \tag{9}$$

For a practical submarine, For a practical submarine $m - Z_{\dot{w}} > 0$, $I_y - M_{\dot{q}} > 0$, and $(m - Z_{\dot{w}})(I_y - M_{\dot{q}}) \gg Z_{\dot{q}}M_{\dot{w}}$ due to the given submarine physical characteristics. Therefore, it is readily established that $b_{11}b_{22} - b_{12}b_{21} \neq 0$, thus B is invertible.

Remark: Note that, physically, the condition $u \neq 0$ and $\theta \neq \pm(2k+l)\pi/2$ means that the submarine forward speed is nonzero and the vehicle is not positioned vertically, which are extreme conditions that do not happen in practice.

2.2 Fault Dynamics

In practical operation sub-system failures may occur. Faults could be stern or bow plane jam, sensor fault, and so forth. Therefore, the dynamic system described in (5) is modified to reflect the fault occurring at time instant the fault occurring at time instant $t \geq T_h$ (4):

$$\ddot{h} = E(z, \theta, u, \dot{z}, \dot{\theta}, \dot{u}) + G(h, v)\beta_h(t - T_h) + Bv \tag{10}$$

where G denotes the unknown dynamics due to fault and function $\beta_h(t-T_h)$ is a time varying defined by:

$$\beta_h(t-T_h) = \begin{cases} 0 & t < T_h \\ 1 - e^{-\mu_h(t-T_h)} & t \geq T_h \end{cases}$$

Function β covers both jump and continuously time-varying faults and $\mu_h > 0$ is a constant reflecting the fault process. For the system to admit a feasible solution under faulty condition, it is needed to assume that there exists a positive constant $\chi > 0$ such that

$$||G(h,v)\beta_h(t-T_h)|| + \chi \leq ||E||.$$

This assumption implies that the fault condition under consideration is within attackable range in that it does not dominate the value of E under normal conditions.

By regrouping, we can rewrite (10) as:

$$\ddot{h} = A(z,\theta,u,\dot{z},\dot{\theta},\dot{u}) + Bv \tag{11}$$

where

$$A(z,\theta,u,\dot{z},\dot{\theta},\dot{u}) = E(z,\theta,u,\dot{z},\dot{\theta},\dot{u}) + G(h,v)\beta_h(t-T_h) \tag{12}$$

It should be noted that the expression for A collects available and uncertain terms. In the later sections, the available part of A is denoted as A_0 and the collection of terms that are not precisely known are denoted as ΔA.

3 Control Design

3.1 Nonlinear Model

First we introduce a generalized variable Sin terms of tracking error as follows:

$$S = \dot{e} + k_0 e$$

where $k_0 > 0$ is a design constant. It is interesting to note that as indicated by S's expression, the tracking error tends to zero asymptotically if $S \to 0$ as $t \to \infty$, since $k_0 > 0$. Therefore, the following development is focused on stabilizing S. With (11), it can be shown that

$$\dot{S} = L(z,\theta,u,\dot{z},\dot{\theta},\dot{u}) + Bv \tag{13}$$

where the lumped term L is defined as

$$L = A(z,\theta,u,\dot{z},\dot{\theta},\dot{u}) - \ddot{h}^{\star} + k_0\dot{e} \tag{14}$$

For a later use, L is regrouped as the sum between the known part $L_0 = A_0 - \ddot{h}^\star + k_0\dot{e}$ and the uncertain part $\Delta L = \Delta A$, here A_0 is the available part of A from (12) and ΔA is the portion of the system that collects all the uncertainties and disturbances.

The objective is to build a control scheme using two neural networks to make h track the desired state $h^\star$. The nonlinear and uncertain nature of $L(\cdot)$ represents the major challenge in control system design. In the preamble of the controller design, the unavailable part of $L(\cdot)$ is reconstructed via a neural network. The reconstruction error and other resulting uncertainties are then estimated using a second NN. As explained, L is reconstructed via NN units as:

$$L_{NN} = L_0 + W_I^T \psi_I \tag{15}$$

where L_0 available part of L, $W_I \in R^{m\times m}$ optimal weight of the NN, and $\psi_I \in R^m$ is the base function of NN (linear function, sigmoid function, hyperbolic tangent function, etc.). Let the discrepancy between L and L_{NN} be denoted by:

$$\epsilon = L - L_{NN} \tag{16}$$

where ϵ is the reconstruction error. The approximation capabilities of the network can be guaranteed only on a subset of the entire plant state space (5; 6). In this development, the following model for the description of the reconstruction error is considered:

$$\epsilon = r\epsilon_1 + (1-r)\epsilon_2 \tag{17}$$

where $||\epsilon_1|| \leq d_0 < \infty$ and

$$||\epsilon_2|| \leq ||L - L_{NN}|| = ||L - L_0 - W_I^T \psi_I|| \equiv d_1 \tag{18}$$

In (17), r is an indicator with 1 indicating that the NN works satisfactorily (reconstruction error is uniformly bounded), and with 0 indicating that the NN fails to work well such that the reconstruction error cannot be bounded by a constant. It should be noted that d_0 and $d_1(\cdot)$ are still unavailable and should not be directly included in the control scheme.

The control scheme is given by:

$$v = B^{-1}(-k_0 S - \hat{L}_{NN} - u_c) \tag{19}$$

where the estimate neural network of the uncertain part:

$$\hat{L}_{NN} = L_0 + \hat{W}_I^T \psi_I \tag{20}$$

is the output of the first NN (6). The terms used in the controller and estimation of the first neural network are: u_c a compensating signal, $k_0 > 0$ a constant, and $\hat{W}_I$, the estimate of the ideal weight values.

3.2 Stability Analysis

The representation of the first NN is shown in Fig. 3, where Ψ_i denotes the basis functions of the neural networks used and W_i the weights associated with them. The generalized variable in (13) can be rewritten:

$$\dot{S} = -k_0 S + \tilde{W}_I^T \psi_I - u_c + \epsilon \tag{21}$$

where $\tilde{W}_I = W_I - \hat{W}_I$ is the weight estimated error viewed as the difference between the ideal weight and the estimated weight, and ϵ is the error to be attenuated by the second NN unit. The function of the second NN unit is to reconstruct α:

$$\hat{\alpha} = \sum_{i=1}^{N} \hat{W}_i \psi_{II_i} = \hat{W}_{II}^T \psi_{II} \tag{22}$$

where $N = N_1 + 2$, $\hat{W}_{II}$ the estimate of the ideal weight values, and Ψ_{II} is a basis function satisfying: $\Psi_{ll_i} \geq Y_i$.

Based on $\hat{\alpha}$, the control's action of the second NN unit is given by (3):

$$u_c = \frac{\sum_{i=1}^{N} \psi_{II_i} \hat{W}_{II}^T \psi_{II}}{\sum_{i=1}^{N} \psi_{II_i} ||S|| + n} S \tag{23}$$

where $n > 0$ is a small number chosen by designer.

Lyapunov's stability theory is used to prove that the proposed new adaptive scheme is able to achieve the control objective. The first step is to achieve the tuning algorithms for $\hat{W}_I$ and $\hat{W}_{II}$ by considering the following Lyapunov candidate function:

$$V = V_0 + V_1 + V_2 \tag{24}$$

with terms in the above equation:

$$V_0 = \frac{1}{2} S^T S \tag{25}$$

$$V_1 = \frac{1}{2g_I} tr(W_I - \hat{W}_I)^T (W_I - \hat{W}_I) \tag{26}$$

$$V_2 = \frac{1}{2g_{II}} (W_{II} - \hat{W}_{II})^T (W_{II} - \hat{W}_{II}) \tag{27}$$

where $g_I > 0$ and $g_{II} > 0$ are free design parameters (3). From the definition of Lyapunov's stability theory, the derivative of (25) – (27) must be negative

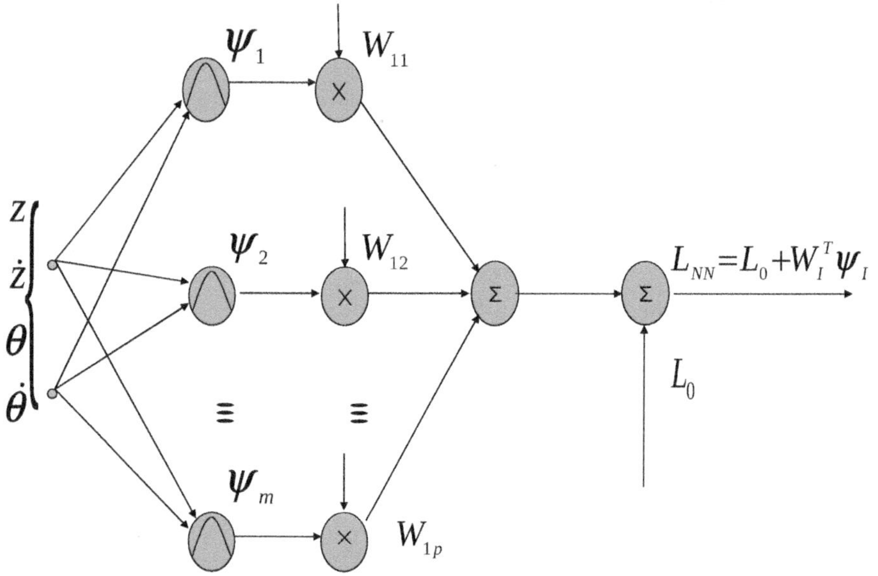

Fig. 3 View of the first NN

in order to achieve the desired system stability. This goal can be achieved by carefully choosing the estimated weights (see detailed proof in (3)):

$$\dot{\hat{W}}_I = -\sigma_1 \hat{W}_I + g_I \psi_I S^T \tag{28}$$

$$\dot{\hat{W}}_{II} = -\sigma_2 \hat{W}_{II} + g_{II} \frac{\sum_{i=1}^{N} \psi_{II_i} ||S^T S|| \psi_{II}}{\sum_{i=1}^{N} \psi_{II_i} ||S|| + n} \tag{29}$$

where $\sigma_I > 0$ and $\sigma + II > 0$ are design parameters chosen to prevent weight drift.

Using the resulting weights it can be seen that $V \to \xi/\lambda_{min}$ as $t \to \infty$, which implies that $||S||$, $||\hat{W}_I||$ and $||\hat{W}_{II}||$ are semi-globally uniformly bounded (6).

4 Simulation Results

In this section, simulations are presented to illustrate the effectiveness of the proposed neuro-adaptive fault-tolerant control in tracking a desired depth and pitch angle of a submarine. Matlab is used for the simulation. The coefficients were collected from the DUPRA SUBOFF model. The simulation models include all non-linearities in the equations of motion, and the values of the parameters and coefficients used are (for more information, see (3)):

$ff = 2, z_{gb} = 1ft;$
$L = 300ft, p = 1.94slug/ft^3;$
$m = 0.0018296 * 0.5 * p * L^3;$
$k_0 = 10, \sigma_1 = \sigma_2 = 0.5, \Delta T = 0.005sec.$

And noise was introduced to the system by $n = \sin(t) + \cos(2t)$.

The basis functions for the first NN unit are chosen as:

$$\psi_I(c_{1i}, X, \alpha) = \frac{1 - e^{-\alpha(||X|| - c_{1i})^2}}{1 + e^{-\alpha(||X|| - c_{1i})^2}} \tag{30}$$

where ($c_{1i} = 0.2, 0.4, \ldots$, $||X|| = [z\ \dot{z}\ \theta\ \dot{\theta}]^T$, $\alpha = 0.5$, $\mu = 0.5$).

The basis functions for the second NN unit are selected as in the following:

$$\psi_{II}(c_{2i}, X, \alpha) = \frac{c_{2i} - c_{2i}e^{-\alpha(||X|| - c_{2i})^2}}{\mu + e^{-\alpha(||X|| - c_{2i})^2}} \tag{31}$$

where ($c_{2i} = 0.3, 0.7 \ldots$).

Ten NN basis functions were used for each of the NN units: $\psi \in R^j$, $j = 10$, and all the weights are initialized at zero. The expressions for fault used in the simulation are:

$$G_z = 100(z^2 + 2\cos\theta)$$
$$G_\theta = 200[z/2 + \sin(2\theta)]$$

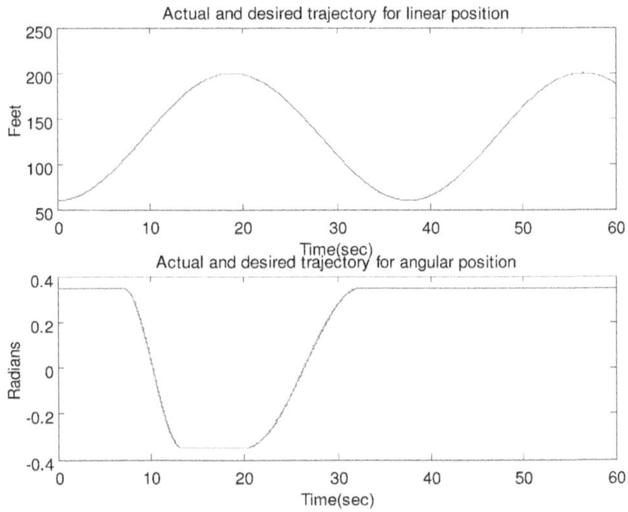

Fig. 4 Tracking when diving from 60 to 200 feet

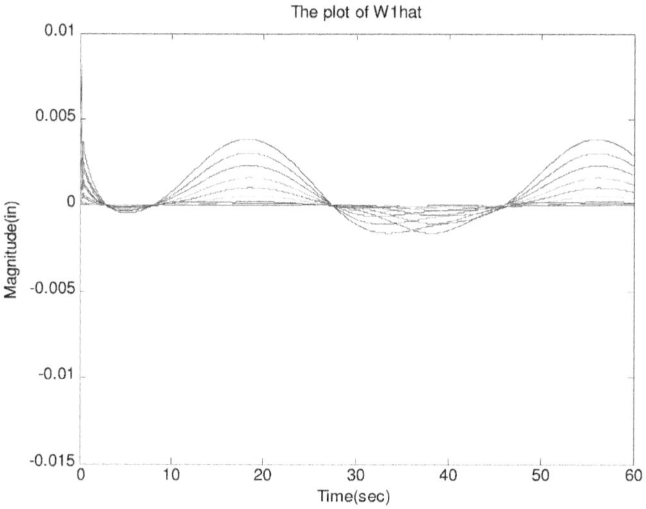

Fig. 5 Estimated weights for the first NN ($\hat{W}_I$)

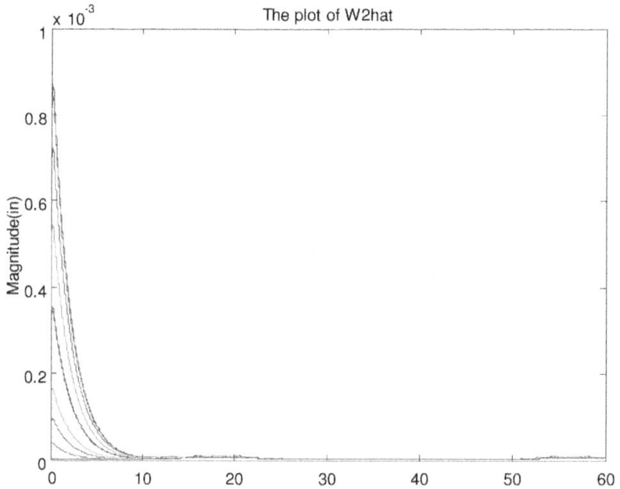

Fig. 6 Estimated weights for the second NN ($\hat{W}_{II}$)

An extreme situation is simulated: the submarine dives from 60 to 200 feet, while a shift in the gravitational center occurs in combination with a varying desired pitch angle, forward speed and fault dynamics. The depth and pitch tracking are only slightly degraded, but error is still very small and within the acceptable range (as seen in Fig. 4, Fig. 5, and Fig. 6). The fault is

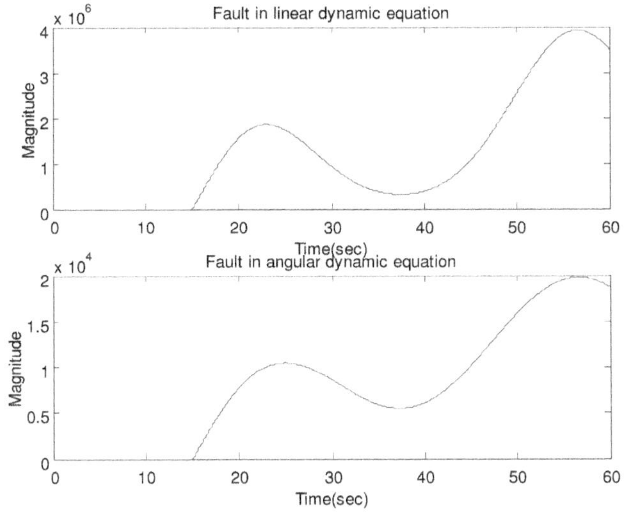

Fig. 7 Dynamic of faults

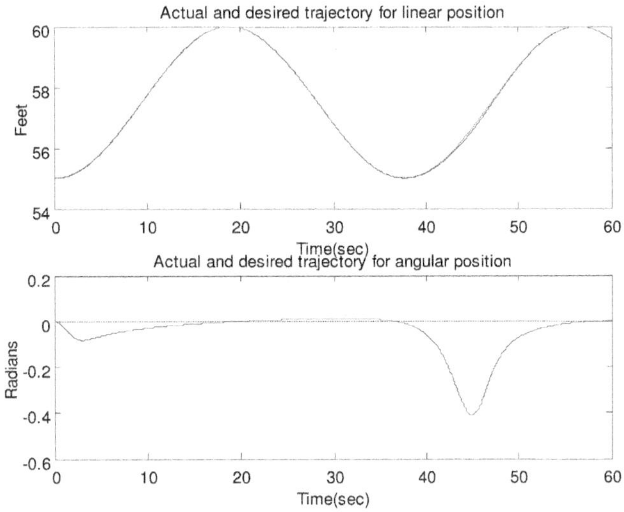

Fig. 8 Tracking under stern plane jam

instantiated at time $T_h = 15$ seconds, and their dynamics are presented in Fig. 7.

The capability to recover from failures is very important to the safety of the crew and the ability to complete the mission assigned. The stern plane is jammed at 0 degrees while the submarine is commanded to dive from 55 feet to 60 feet and then rise back to 55 feet. The speed of the submarine is

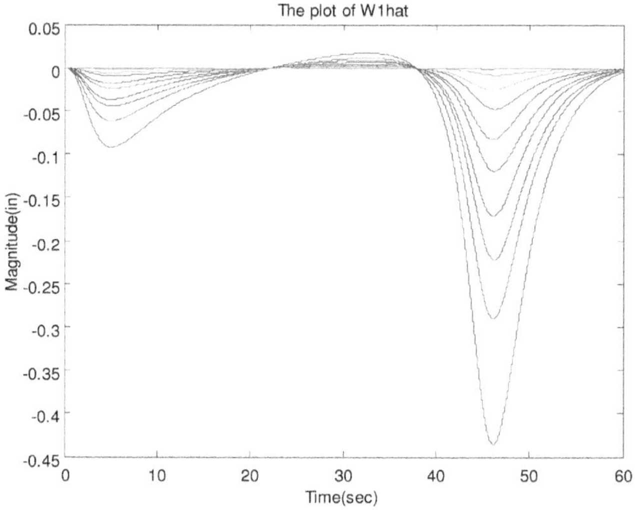

Fig. 9 Estimated weights for the first NN ($\hat{W}_I$)

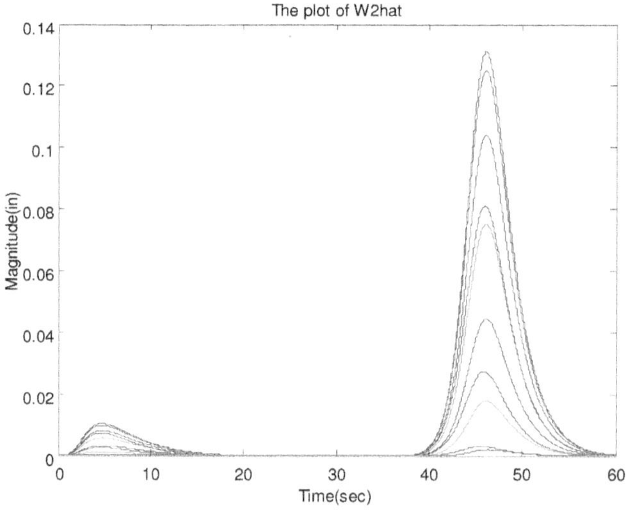

Fig. 10 Estimated weights for the second NN ($\hat{W}_{II}$)

considered time variant, and a shift in the gravitational center is considered in combination with an occurrence of fault. The plots are presented in Fig. 8, Fig. 9, and Fig. 10. The depth and pitch tracking are only slightly degraded, but error is still very small and within the acceptable range. The fault is instantiated at time $T_h = 15$ seconds.

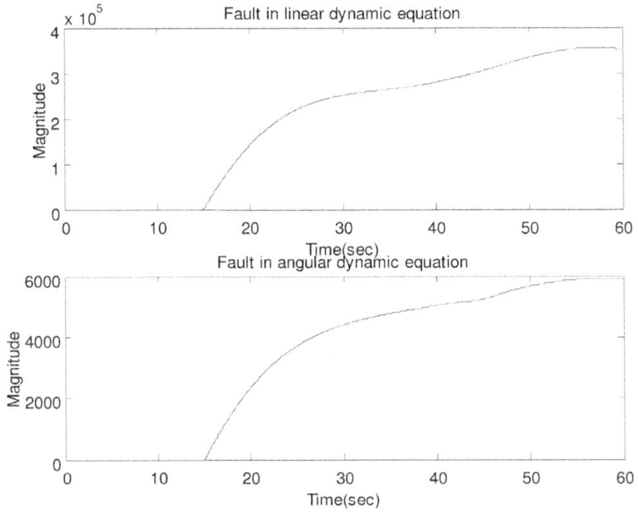

Fig. 11 Dynamic of faults

5 Conclusions

This chapter presented a highly model-independent neuro-adaptive control scheme for depth and the pitch angle control of submarines operating in shallow waters. The control scheme does not involve the bounds on ideal weights, reconstruction error, or the nonlinear function being approximated. This means that a preliminary analysis of the system and the structure of the NN is not needed. The salient feature of the proposed method also lies in its flexibility and simplicity in design and implementation.

References

1. Tolliver, J.V.: Studies of Submarine Control for Periscope Depth Operations, M.S. Thesis, Naval Postgraduate School (1996)
2. Papoulias, F.A., Bateman, C.A., Ornek, S.: Dynamic Loss of Stability in Depth control of Submersible Vehicles. Applied Ocean Research 17, 205–216 (1995)
3. Song, Y.D., Medorian, D.G.: Control of Submarine Operating in Shallow Waters. Technical Report, Center for Cooperative Systems, M.S. Thesis, North Carolina A&T State University (2002)
4. Song, D., Xue, X.: Fault-tolerant control of composite systems with faulty dynamics, UNSW, Sydney, Australia (2002)
5. Hunt, J., Sbarbaro, D., Zbikowsji, R., Gawthrop, P.J.: Neural Networks for Control Systems – a survey. Automatica 26, 1085–1112 (1992)
6. Song, Y.D.: Neuro-adaptive control with application to robotic systems. Journal of Robotics Systems 14(6), 433–447 (1997)

7. Bacon, D.K.: Integration of a Submarine into NPSNET, M.S. Thesis, Naval Postgraduate School (1995)
8. Lin, C., Xu, R., Kwan, C., Haynes, L.: Submarine Pitch and Depth Control Using FMAC Neural Network. In: Proc. of Am. Control Conference (1998)
9. Brutzman, D.: A Virtual World for an Autonomous Underwater Vehicle. PhD Dissertation, Naval Postgraduate School (2004)
10. Papoulias, F.: Nonlinear Studies of Dynamic Stability of Submarines in the Dive Plane. Journal of Ship Research 39(4), 347–356 (1995)
11. `http://blog.uncovering.org/en/archives/2008/05/phoenix_1000_a_luxurious_submarine.html`
12. Demirci, U., Kerestecioglu, F.: Active reconfigurable control of a submarine with indirect adaptive control. In: Proc. of OCEANS, vol. 5, pp. 2491–2495 (2003)
13. Naik, M.S., Singh, S.N., Mittal, R.: Biologically-Inspired Adaptive Pectoral-Like Fin Control System For CFD Parameterized AUV. In: Proc. of Underwater Technology and Workshop on Scientific Use of Submarine Cables and Related Technologies, pp. 371–377 (2007)
14. Sheng, L., Xiuli, Z., Yanchun, D., Wanlong, L., Meiyan, T.: Research of adaptive fuzzy controller in pitch regulation system of ocean submarine. In: Proc. of Mechatronics and Automation, vol. 2, pp. 847–852 (2005)

Stick-Slip Friction Compensation Using a General Purpose Neuro-Adaptive Controller with Guaranteed Stability

Ali Reza Mehrabian and Mohammad Bagher Menhaj

Abstract. A general purpose neuro-adaptive controller, which realizes an indirect-adaptive control strategy, is introduced. The proposed algorithm is based on the use of two Multi-Layer feed-forward Perceptron (MLP) Neural Networks (NNs), which are trained using a momentum back-propagation (MBP) algorithm. One of the MLP NNs is used to identify the process. The other MLP NN is used to generate the control signal based on the data provided by the NN identifier. Training is done on-line to tune the parameters of the neuro-identifier and neuro-controller that provides the control signal. Pre-learning is not required and the structure of the overall system is very simple and straightforward, no additional controller or adaptive signal is needed. Tracking performance is guaranteed via Lyapunov stability analysis, so that both tracking error and neural network weights remain bounded. An interesting fact about the proposed approach is that it does not require a NN being capable of globally reconstructing the nonlinear model.

Several simulation examples are reported to demonstrate the merits of the proposed algorithm. As is shown in the simulations, the developed control algorithm can deal with different types of challenges that might happen in real-time applications, including the change of the reference model and the effect of applied unknown disturbances. The application of the proposed neuro-control algorithm to the adaptive control of electro-mechanical systems subject to stick-slip friction is shown in the last section of this paper. Reported simulations reveal that the proposed algorithm is able to eliminate the effect of this nonlinear phenomenon on the performance of the system.

1 Introduction

The design of control algorithms without considering friction in machineries causes steady-state tracking errors and/or oscillations in the closed-loop system's response. Friction causes heat and fatigue in machines, which reduces

J. Schumann, Y. Liu (Eds.): Appl. of Neural Networks in High Assur. Sys., SCI 268, pp. 179–203.
springerlink.com

their performance and operational life. During the last two decades, the design of control laws for friction compensation has been an interesting research area in the control community; i.e. *Dupont* has developed a PD controller for avoiding the friction with unknown parameters (1). Several research studies have been carried out to develop adaptive/nonlinear or neuro-adaptive compensators for simplified models of the friction (2)-(7). A complex model of stick-slip friction was introduced in (8), which reproduces the friction during sticking and slipping motions as well as the "Stribeck" effect (9). A more complicated friction model was developed by *Canudas de Wit et al.* (10) by capturing dynamic friction effects like the "Stribeck" effect, hysteresis and the stick-slip limit cycle. Different control approaches for friction compensation, e.g. adaptive-sliding controller (11) and neural-network-based control (12), have been proposed based on the model developed in (10). It is worth mentioning that the model presented in (8), which captures all three major components of frictions namely Coulomb force, viscous force and Stribeck force, was used to develop different controllers for friction compensation by many scientists (9), (13)-(15).

The problem of control is some kind of decision making; given the observation of the state of the process it has to be decided from encoded knowledge what action to take. Neural networks (NNs) with their massive parallelism and ability to learn any kind of nonlinearity are a favorable methods. A neuro-controller, in general, performs a specific form of adaptive control with the controller being in the form of a multilayer NN and the adaptable parameters being defined as the adjustable weights.

Among the different aspects of a controller, its ability to stabilize the overall system is the most important in real world problems. Recently, there has been some research concerning this subject. In fact, there is a rich and seminal theory on dynamic behavior of NNs, e.g. see *Cohen* and *Grossberg* (16), *Hopfield* (17), *Kosko* (18), and *Guez* (19), as well as on stability and the number of equilibrium points of recurrent NNs, e.g. see *Kelly* (20), *Vidyasagar* (21), and *Jim et al.* (22). In contrast, there is only limited work on stability of NNs as controllers for dynamic systems. *Tzirkel-Hancock* and *Fallside* (23) developed two neural based estimators to estimate output derivatives with respect to inputs and states for feedback linearizable non-linear systems. They showed that, under some specified conditions, like exact estimation of derivatives, tracking error will approach zero.

Among the pioneering papers are the studies done by *Lewis et al.* (24), (25). They applied a NN accompanied with a PD controller or a robustifying term to a robot system. They proposed a NN to cancel out the non-linear part of the robot. They used special properties concerning inertia and friction matrices, and showed that it is possible to use a modified BP-like algorithm to train the NN to reduce the tracking error to zero. This work was extended and modified in *Kwan et al.* (26) for electrically driven robots. *Kuntanapreeda* and *Fullmer* (27), and *Park* and *Park* (28) applied directly a single hidden layer NN controller to a general, locally stable non-linear system. They gave some implicitly defined conditions on the weight matrix of the NN in order to make the overall system locally stable. *Sadegh* (29) introduced a method to design adaptive NN controllers by local stability analysis. *Polycarpou* (30) has pointed out that some of the conditions commonly considered in adaptive NN control design can be relaxed for special cases. In *Chen* and *Khalil* (31), (32), and *Chen* and *Liu* (33), two NNs are used to estimate the nonlinearity of a feedback linearizable system. The stability results are locally valid in the parameter space, so that the NNs require off-line training. *Fabri* and *Kadirkamanathan* (34), and *Sun et al.* (35) used radial basis function NNs together with a sliding mode controller to adaptively stabilize a non-linear model. *Jagannathan et al.* (36), (37) have developed NN-based controllers for a special class of non-linear discrete systems. In addition to these studies, we should point out the work by *Suykens* and *Vandewalle* (38). Their results concern conditions on the weight matrix of a NN so that the overall system consisting of a NN controller and a NN model of the plant becomes stable. Their design procedures are mainly based on complicated matrix equalities or inequalities.

In this study, we first present a new general purpose model reference adaptive control scheme by NNs based on the momentum back-propagation (MBP) learning rule to adapt parameters of the NN controller. The control scheme consists of a neuro-identifier as well as a neurao-controller. Training is done online and no pre-learning is required. The structure of the overall system is very simple and straightforward. The proposed neuro-control scheme has the following features: first of all, no additional controller or robustifying signal is required, while the tracking performance of the system is guaranteed via a Lyapunov stability analysis. It is shown that when the two adjustable sets of parameters of the control system are tuned properly, both tracking error and the NN weights remain bounded while the closed-loop system remains stable and shows a satisfactory response. An interesting fact about the approach is that it does not require a NN being capable of reconstructing a global model for the non-linear plant. We then present the development and application of the neuro-control scheme for unknown nonlinear stick-slip friction compensation based on the model developed in (8) for the regulation of a DC-motor driven nonlinear mechanical system with friction. Simulation results are presented to show the merits of the control scheme.

This study is organized as follows. We begin with description of an indirect control strategy, NN structure and identification and control scheme using NNs in Sect. 2. In Sect. 3, we present the stability analysis of the proposed control method. Sect. 4 presents an implementation of the proposed control approach using SIMULINK[1] and introduces fully the structures and blocks of the system. Sect. 5 is devoted to present some illustrative examples and

[1] `http://www.mathworks.com`

to discuss the simulation results of the proposed scheme for controlling nonlinear dynamical systems. The development and application of the introduced neuro-control approach for compensating the stick-slip friction is discussed in Sect. 6. Sect. 7 concludes the study.

2 The Neural-Network-Based Control Strategy

2.1 Indirect Adaptive Neuro-Controller

NNs can be used to adaptively control a nonlinear system in the presence of uncertainty. Generally speaking, direct and indirect adaptive control schemes represent two distinct methods for the design of adaptive controllers. To use neural computations to design adaptive controllers, we may use Direct Adaptive Neuro-Control (DANC) and Indirect Adaptive Neuro-Control (IANC) schemes. In the DANC the parameters of the controller are directly adjusted to minimize the tracking error, while in the IANC scheme, parameters of the plant under study are estimated using a NN, called the identifier and based on these estimates, the controller parameters are then adjusted (39). The latter scheme is used in this paper (for more details on the proposed control scheme see (40)).

Fig. 1 represents a schematic diagram of a modified indirect serial-parallel adaptive neuro-controller connected to a plant. It basically consists of a MLP based NN used as a neuro-controller (NNC), along with a second NN used for identifying (NNI) the plant. This IANC scheme incorporates a stable reference model that reflects the desired closed-loop dynamics of the plant. The identifier is designed to track the plant response online by minimizing the identification error ($\mathbf{e}_i$), while the controller is designed to generate control variables such that the output of the plant tracks the reference model for a given bounded reference input. The latter is achieved by adjusting parameters of the controller via an adjustment mechanism so as to minimize the error between the outputs of the reference model and the plant ($\boldsymbol{\epsilon}$).

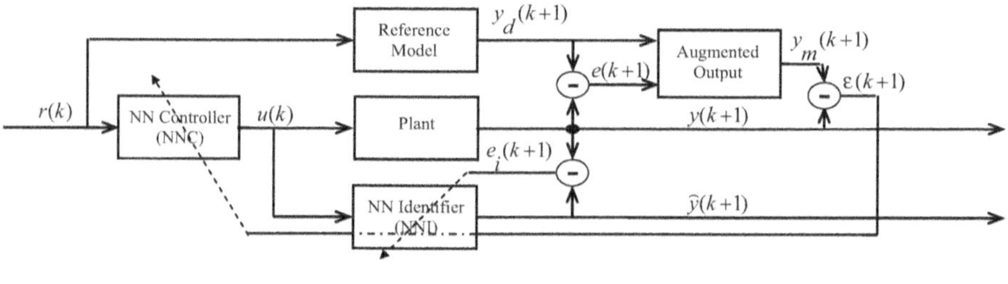

Fig. 1 The proposed IANC scheme

2.2 Neural Network Scheme

A two layer feedforward NN of m tangent-sigmoid function neurons having n inputs in full detail is shown in Fig. 2. A network can have several layers. Each layer has a weight matrix $\mathbf{W}$, a bias vector $\mathbf{b}$, and an output vector, $\mathbf{u}$.

Feedforward networks often have one or more hidden layers of *sigmoid* or *tangent-sigmoid* neurons followed by an output layer of linear neurons. Multiple layers of neurons with nonlinear transfer functions allow the network to learn nonlinear and linear relationships between input and output vectors. The linear output layer lets the network produce values outside the range -1 to $+1$. For multiple-layer networks we use the number of the layers to determine the superscript on the weight matrices and biases, or just mention first layer (input layer) weights by $\mathbf{W^1}$, and the second layer weight by $\mathbf{W^2}$.

2.3 Control Oriented On-Line Identification Method

The first step in the control design process is the development of the plant model. The performance of the NNC is highly dependent on the accuracy of the plant identification. There are generally two methods for development of the neural model of a system. The first method, which is easier in computation is off-line training, where weights of the NNI are adjusted off-line using preexistent data from the plant's input and output response. Using this set of data, the NNI can be trained using the elementary back-propagation (BP) algorithm (good for pattern-by-pattern training) or batch training methods like the Marquardt algorithm (41). One drawback of this method is its dependency on training data. Training data which has been collected off-line should have some specific properties, for example, we need to ensure that the plant input is sufficiently exiting (42), and we further need to be sure that the system inputs and outputs cover the operation range. If we consider these points and assume that the identification is performed well, the question is whether the non-linear system dynamics remains unchanged through time or not. It is very well known that the identification is not perfect, and besides, the characteristics of a dynamic system vary through time, such that an off-line identification method is left unprotected against these changes; this is the main disadvantage of the off-line method.

In contrast, if the identification is performed on-line we will have a massive volume of on-line computation (a lengthy time-consuming task). The control strategy designer must find tradeoff between the increased robustness and adaptability of on-line training and the low computational complexity of the off-line methodology.

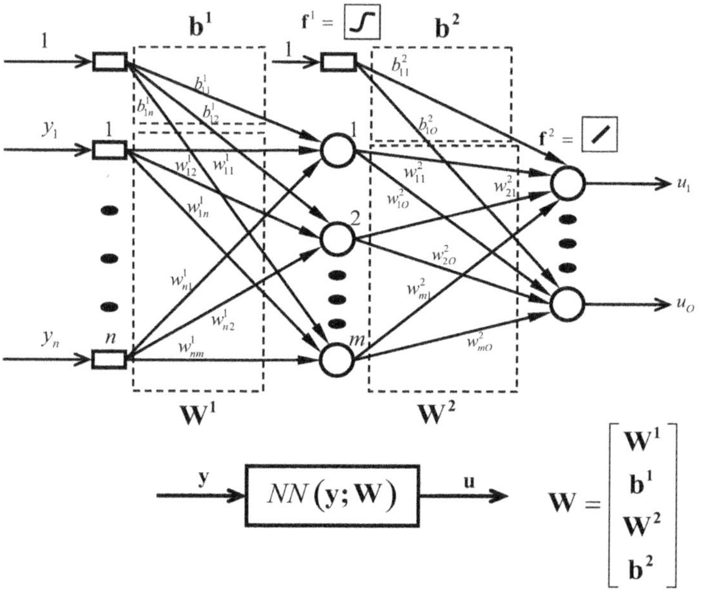

Fig. 2 Schematic of a two layer feedforward NN: Each ○ represents a NN function and □ represents an input to the NN

2.4 Mathematical Description of the Control Scheme

Consider a plant governed by the following nonlinear difference equation:

$$\begin{aligned} \boldsymbol{y}_p(k+1) &= f(\mathbf{Y}_p^{k,l}, \mathbf{u}^{k,q}) \\ \boldsymbol{y}_p(0) &= \boldsymbol{y}_{p0}, \qquad\qquad \forall k \in \mathbb{N}, \end{aligned} \tag{1}$$

where $\mathbf{u}$ is the plant output, k is the time index and $\mathbf{X}^{k,l}$ means:

$$\mathbf{Y}^{k,l} = [\boldsymbol{y}(k), \boldsymbol{y}(k-1), \cdots, \boldsymbol{y}(k-l+1)],$$

and $\boldsymbol{y}_p(k) \in \mathbb{R}^n$ is the output vector, $\boldsymbol{u}(k) \in \mathbb{R}^r$ is the control vector, f is a smooth nonlinear function, $\boldsymbol{y}_p(0)$ is an initial output vector, k is the time index, $\mathbb{N}$ is the set of natural numbers and m and l are the number of delayed values of plant input and plant output respectively.

Consider a stable reference model generating the desired response governed by:

$$\begin{aligned} \boldsymbol{y}_d(k+1) &= f_d(\mathbf{Y}_d^{k,g}, \mathbf{r}^{k,s}) \\ \boldsymbol{y}_d(0) &= \boldsymbol{y}_{d0}, \qquad\qquad \forall k \in \mathbb{N}, \end{aligned} \tag{2}$$

where $\boldsymbol{y}_d(k) \in \mathbb{R}^n$ is the reference model output vector, $\boldsymbol{r}(k) \in \mathbb{R}^n$ is the reference input, f_d is usually a linear function and $\boldsymbol{y}_d(0)$ is a given initial output vector for the reference model.

The control strategy is to find a feasible control input with the following general structure to establish the closed-loop stable system whose response tracks the desired trajectory:

$$\boldsymbol{u}(k) = \mathrm{NN}_c(\mathbf{Y}_p^{k,l}, \mathbf{u}^{k,q}, \mathbf{r}^{k,s}; \mathbf{W}_c) \tag{3}$$

where NN_c is a MLP based NN parameterized by the weights $\mathbf{W}_c$ presented below:

$$\mathbf{W}_c = \left[\mathbf{W}^1, \mathbf{b}^1, \mathbf{W}^2, \mathbf{b}^2\right]^T$$

such that the corresponding plant output is nearest, in some norm to the augmented output $\boldsymbol{y}_m(k+1)$ defined below. This signal which is a modified version of the desired signal is to be specified a priori as a function of the plant output and the reference model input. An appropriate model to generate $\boldsymbol{y}_m(k+1)$ is:

$$\begin{aligned} \boldsymbol{y}_m(k+1) =& \boldsymbol{y}_d(k+1) - \mathbf{A}(\boldsymbol{y}_d(k) - \boldsymbol{y}(k)) \\ & - \mathbf{B}(\boldsymbol{y}_d(k-1) - \boldsymbol{y}(k-1)) \end{aligned} \tag{4}$$

where $-\mathbf{A}$ and $-\mathbf{B}$ are Hurwitz matrices. The "Augmented Output" block in Fig. 1 implements (4). This signal (instead of the original desired signal) will be used to tune the parameters of the neuro-controller. In order to see why we are using the augmented signal, (4) is rewritten as:

$$\boldsymbol{y}_m(k+1) = \boldsymbol{y}_d(k+1) - \mathbf{A}\boldsymbol{e}(k) - \mathbf{B}\boldsymbol{e}(k-1) \tag{5}$$

where $\boldsymbol{e}(k) = \boldsymbol{y}_d(k) - \boldsymbol{y}(k)$. The desired signal, $\boldsymbol{y}_d$, in general is obtained from a reference model preferably linear, time invariant system, like:

$$\begin{aligned} \boldsymbol{x}_d(k+1) &= \mathbf{A}_d\boldsymbol{x}_d(k) + \mathbf{B}_d\boldsymbol{r}(k) \\ \boldsymbol{y}_d(k) &= \mathbf{C}_d\boldsymbol{x}_d(k) \end{aligned} \tag{6}$$

in which $\mathbf{A}_d$ represents an asymptotically stable matrix.

Now consider the following performance index:

$$\begin{aligned} \hat{F}_c(\boldsymbol{x}) =& \left[\boldsymbol{y}_m(k+1) - \boldsymbol{y}(k+1)\right]^T \left[\boldsymbol{y}_m(k+1) - \boldsymbol{y}(k+1)\right] \\ =& \boldsymbol{\epsilon}^T(k+1)\boldsymbol{\epsilon}(k+1) \end{aligned} \tag{7}$$

A suitable control output, $\boldsymbol{u}(k)$, is generated by adjusting the parameters $\mathbf{W}_c$ of the neuro-controller NN_c through minimization of (7).

2.5 Training Multilayer Neural Network (MLP)

Now that we know the control strategy, the next step is to determine a procedure for selecting the NNC parameters (weights and biases). The networks used in this paper are Multilayer Perceptron (MLP) type NNs having a single hidden layer with *tangent-sigmoid* function and the *momentum back-propagation* algorithm (44) for training. It is shown that two-layer networks, with tangent-sigmoid transfer functions in the hidden layer and linear transfer functions in the output layer, can approximate virtually any function of interest to any degree of accuracy, provided that there exist sufficiently many hidden units (43). The parameter update rule for momentum back-propagation algorithm is:

$$\begin{aligned} \Delta\mathbf{W}^m(k) &= \gamma\Delta\mathbf{W}^m(k-1) - (1-\gamma)\alpha\boldsymbol{s}^m(\boldsymbol{a}^{m-1})^T \\ \Delta\mathbf{b}^m(k) &= \gamma\Delta\mathbf{b}^m(k-1) - (1-\gamma)\alpha\boldsymbol{s}^m \end{aligned} \tag{8}$$

where Δ represents update (correction term), γ is the *momentum term* and α is the *learning rate*. $\boldsymbol{s}^m$ is sensitivity of the layer m is defined as:

$$s_i^m = \frac{\partial \hat{F}_c}{\partial n_i^m} \tag{9}$$

which measures the sensitivity of the cost function $\hat{F}_c$ to the changes in the ith element of the NN input at layer m (see (44) for more details). It is shown in (44) that derivative of cost function with respect to network weights is found by:

$$\frac{\partial \hat{F}_c}{\partial w_{i,j}^m} = s_i^m a_j^{m-1}, \quad \frac{\partial \hat{F}}{\partial b_i^m} = s_i^m \tag{10}$$

where $x_{i,j}^m$ presents the ith row and jth column of mth layer's weight or bias; $\boldsymbol{a}_j^m$ is jth column of the output of the NN's mth layer.

Remark 1. Back-propagation NNs require that all training targets be normalized between 0 and 1 for training. This is because an output node's signal is restricted to a 0 to 1 range.

2.6 Back-Propagation through the Model

In this step, we use the NN model (NNI) to back-propagate the controller error ($\boldsymbol{\epsilon}$) back to the NN controller (NNC). The NNC then learns the desired control command through back-propagation algorithm. This can be formulated as follows:

$$\frac{\partial \hat{\mathbf{F}}_c(k+1)}{\partial \mathbf{W}_c(k)} = \frac{\partial \hat{\mathbf{F}}_c(k+1)}{\partial \boldsymbol{u}(k)} \frac{\partial \boldsymbol{u}(k)}{\partial \mathbf{W}_c(k)} \tag{11}$$

Knowing the fact that plant Jacobian matrix $\boldsymbol{J}(k) = \frac{\boldsymbol{y}(k+1)}{\boldsymbol{u}(k)}$ cannot be assumed to be available a priori, we might use its estimates obtained from $\frac{\hat{\boldsymbol{y}}(k+1)}{\boldsymbol{u}(k)}$ back-propagated through the NN plant-identifier (NNI).

3 Stability Analysis

This section presents a detailed proof for asymptotic behavior of the neuro-controller that makes the overall closed-loop systems to track the desired trajectory. To do so, we first rewrite the tracking error dynamics in terms of $\boldsymbol{\epsilon}$, then propose a Lyapunov function. Knowing that $\boldsymbol{\epsilon}(k) = \boldsymbol{y}_m(k) - \boldsymbol{y}(k)$, (5) becomes:

$$\boldsymbol{e}(k+1) = \boldsymbol{\epsilon}(k+1) + \mathbf{A}\boldsymbol{e}(k) + \mathbf{B}\boldsymbol{e}(k-1) \tag{12}$$

where $\boldsymbol{e}(k) = \boldsymbol{y}_d(k) - \boldsymbol{y}(k)$. Let us use the following frequently used function, as a candidate Lyapunov function:

$$V(\boldsymbol{e}(k)) = \boldsymbol{e}^T(k)\boldsymbol{e}(k) \tag{13}$$

Now we proceed as follows:

1. The function V is trivially positive definite;
2. We do need to show the function is non-increasing. This can be done as:

$$\begin{aligned}
\Delta V(\boldsymbol{e}(k)) =& V(\boldsymbol{e}(k+1)) - V(\boldsymbol{e}(k)) \\
=& \boldsymbol{e}^2(k+1) - \boldsymbol{e}^2(k) \\
=& [\boldsymbol{\epsilon}(k+1) + \mathbf{A}\boldsymbol{e}(k) + \mathbf{B}\boldsymbol{e}(k-1)]^T \\
& \times \boldsymbol{\epsilon}(k+1) + \mathbf{A}\boldsymbol{e}(k) + \mathbf{B}\boldsymbol{e}(k-1) - \boldsymbol{e}^2(k) \\
=& \boldsymbol{\epsilon}^2(k+1) + \boldsymbol{\epsilon}^T(k+1)\mathbf{A}\boldsymbol{e}(k) \\
& + \boldsymbol{\epsilon}^T(k+1)\mathbf{B}\boldsymbol{e}(k-1) + \boldsymbol{e}^T(k)\mathbf{A}^T\boldsymbol{\epsilon}(k+1) \\
& + \boldsymbol{e}^2(k)\mathbf{A}^2 + \boldsymbol{e}^T(k)\mathbf{A}^T\mathbf{B}\boldsymbol{e}(k-1) \\
& + \boldsymbol{e}^T(k-1)\mathbf{B}^T\boldsymbol{e}(k+1) \\
& + \boldsymbol{e}^T(k-1)\mathbf{B}^T\mathbf{A}\boldsymbol{e}(k) \\
& + \boldsymbol{e}^2(k-1)\mathbf{B}^2 - \boldsymbol{e}^2(k) \\
=& \boldsymbol{e}^T(k)(\mathbf{A}^2 - \mathbf{I})\boldsymbol{e}(k) + 2\boldsymbol{\epsilon}^T(k+1)\mathbf{A}\boldsymbol{e}(k) \\
& + 2\boldsymbol{\epsilon}^T(k+1)\mathbf{B}\boldsymbol{e}(k-1) \\
& + 2\boldsymbol{e}^T(k)\mathbf{A}^T\mathbf{B}\boldsymbol{e}(k-1) \\
& + \boldsymbol{e}^T(k-1)\mathbf{B}^T\mathbf{B}\boldsymbol{e}(k-1) \\
& + \boldsymbol{\epsilon}^T(k+1)\boldsymbol{\epsilon}(k+1)
\end{aligned} \tag{14}$$

if

$$\begin{aligned} \boldsymbol{e}^T(k)\mathbf{Q}\boldsymbol{e}(k) >& 2\boldsymbol{\epsilon}^T(k+1)\mathbf{A}\boldsymbol{e}(k) \\ &+ 2\boldsymbol{\epsilon}^T(k+1)\mathbf{B}\boldsymbol{e}(k-1) \\ &+ 2\boldsymbol{e}^T(k+1)\mathbf{A}^T\mathbf{B}\boldsymbol{e}(k-1) \\ &+ \boldsymbol{e}^2(k-1)\mathbf{B}^2 + \boldsymbol{\epsilon}^2(k+1) \end{aligned} \tag{15}$$

where $\mathbf{Q} = -(\mathbf{A}^2 - \mathbf{I})$, then:

$$\Delta V(\boldsymbol{e}(k)) < 0 \tag{16}$$

Thus $\boldsymbol{e}(k) \to 0$ if $\boldsymbol{\epsilon}(k+1)$ satisfies the inequality (16). By proper selection of $\mathbf{A}$ and $\mathbf{B}$, we can make $\boldsymbol{\epsilon}$ decay faster than $\boldsymbol{e}$; this in turn guarantees inequality (16) to hold. Due to our vast simulation studies and experiments, we came up with the following trend regarding how to select $\mathbf{A}$ and $\mathbf{B}$; this mainly depends on the dynamics of the plant. For a system with fast dynamics and low degree of stability or in the presence of severe noise and disturbances, it is recommended to set relatively larger values for $\mathbf{A}$ and $\mathbf{B}$ parameters, while for a system with slower dynamics, small values for $\mathbf{A}$ and $\mathbf{B}$ will work better. The reason behind this is that larger values for $\mathbf{A}$ and $\mathbf{B}$ make the controller more sensitive to changes in the error signal; thus, the controller responds more quickly to the changes and will capture the rapid dynamics of the system. It should be noted that large values of these two matrices may cause high-frequency oscillations in some cases. In general after a few trial and error one can end up with proper values of $\mathbf{A}$ and $\mathbf{B}$ within the unit interval.

4 Implementing the Proposed Adaptive-Neuro Control Method

Implementation of the proposed control algorithm is done using SIMULINK. The control loop block diagram in SIMULINK is shown in Fig. 3. The proposed model mainly consists of six blocks. The first block is the *Reference model* block. The second is the *plant* block, which contains the dynamics of the plant. In reality, this block will be replaced by an unknown plant. The third and fourth blocks are *NN identifier* (model) and *controller*, respectively. The next block is the augmented output builder which implements (4), and the last block is the *controller error sensitivity feedback* block, which is used to find $\frac{\partial \hat{\mathbf{F}}_c(k+1)}{\partial \boldsymbol{u}(k)}$.

4.1 NN Identifier Block

The *NN identifier* block is shown in Fig. 4. As discussed before, the identifier consists of two parts. The first part represents NN structure and the second

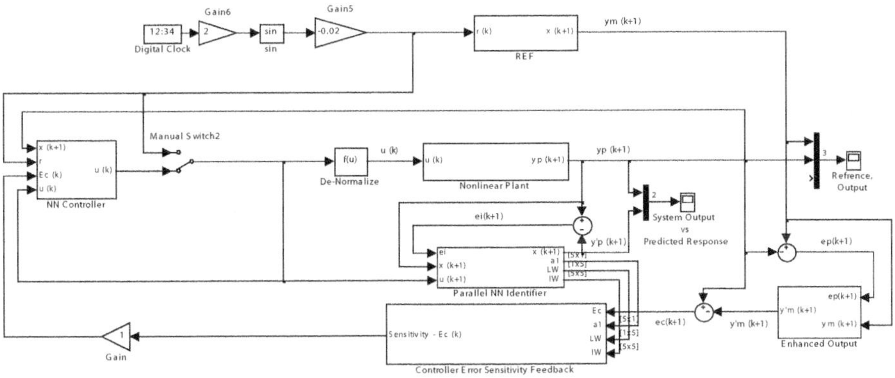

Fig. 3 SIMULINK blocks for the developed control algorithm

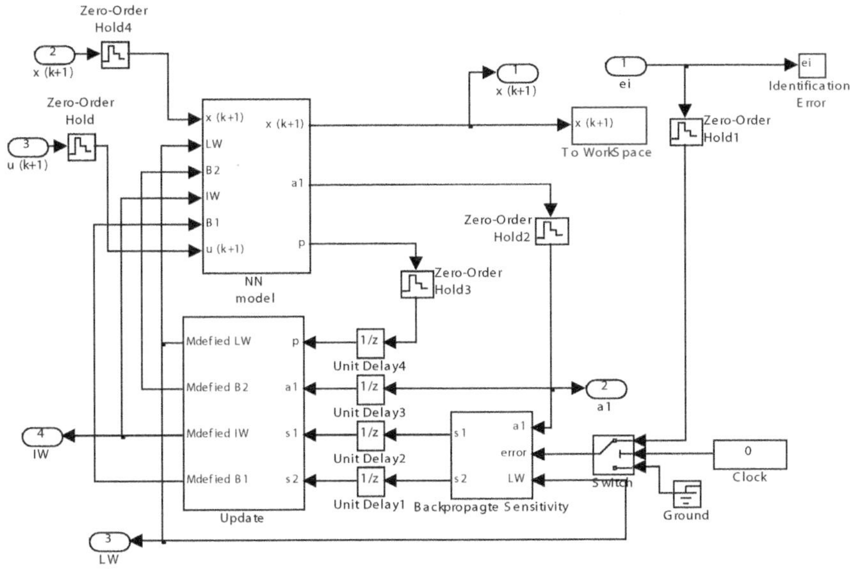

Fig. 4 SIMULINK model of the NN identifier

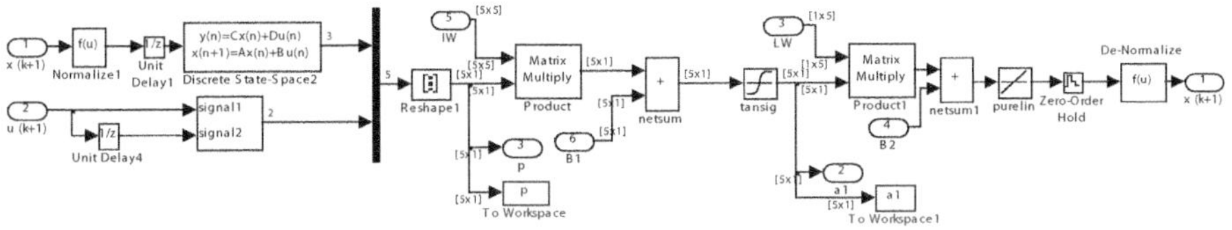

Fig. 5 NN model of the identifier in SIMULINK

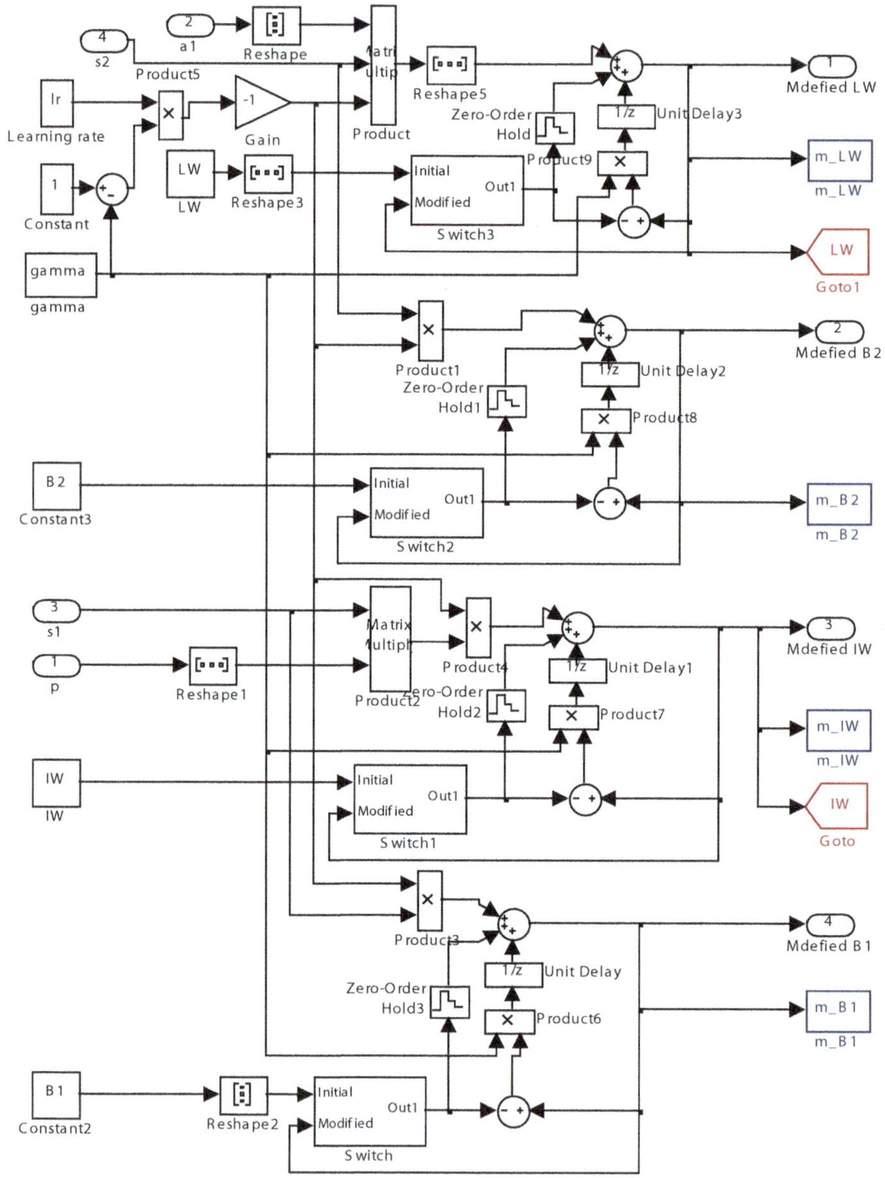

Fig. 6 Updating mechanism of the NN identifier

part is the updating mechanism. The NN model of the identifier is shown in Fig. 5. Block diagram of the updating mechanism is given in Fig. 6. The updating mechanism has capability of implementing MBP algorithm for adaptation.

4.2 NN Controller Block

Neuro-controller block is shown in Fig. 7. Its updating mechanism is the same as the identifier, plus it has the ability of changing learning rate proportional to the tracking or control error ($\boldsymbol{e}$ or $\boldsymbol{\epsilon}$).

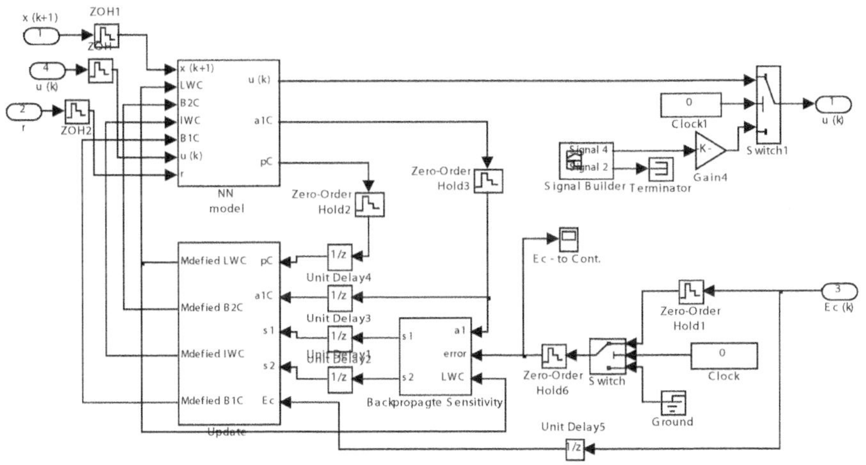

Fig. 7 A model for the NN controller in SIMULINK

4.3 Controller Error Sensitivity Feedback Block

The structure of the *controller error sensitivity feedback* block is shown in Fig. 8. As it was mentioned before, this block is used to find the modified tracking error sensitivity to plant input variations. To do so, identifier weights and the error are fed into this block whose output measures the sensitivity.

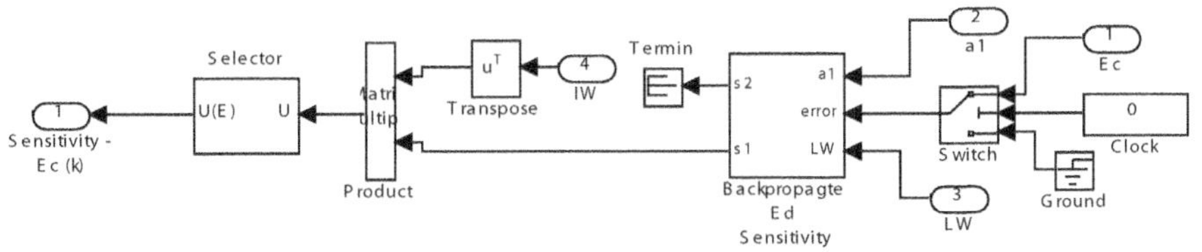

Fig. 8 SIMULINK model of the controller error sensitivity feedback block

5 Simulation Studies

In this section, we perform a few simulations to investigate the adaptation and the robustness of the proposed NN-based control scheme. Before we describe the simulations, the following remarks are worth mentioning:

Remark 2. As mentioned in Sect. 2, the control scheme consists of a NNI and a NNC; so, the identification process is necessary for finding the proper control signal. Due to this necessity the NNI first learns the unknown dynamic system by applying random control signals from the controller, i.e. in the cases we considered in the paper it takes only a few seconds (about 0.2 sec) and then the neuro-controller is put into the loop.

5.1 Example 1: A Non-linear System with a Second-Order Difference Equation and Variable Reference Model

In the first example, we want to show the system's ability in real-time identification and control of an unknown plant. We consider a plant in the following form taken from (39):

$$y(k+1) = f[y(k), y(k-1)] + u(k) \tag{17}$$

where the function:

$$f[y(k), y(k-1)] = \frac{y(k)y(k-1)[y(k)+2.5]}{1+y(k)^2+y(k-1)^2} \tag{18}$$

is assumed to be unknown. The first reference model is described by a second-order difference equation defined as:

$$y_d(k+1) = 0.2y_d(k) + 0.2y_d(k-1) + r(k), \tag{19}$$

and the second one is a first-order difference equation given below:

$$y_d(k+1) = 0.5y_d(k) + r(k), \tag{20}$$

where (19) is applied for $t < 8$ seconds, and (20) is applied for $t \geq 8$ seconds. In practice, switching to different reference models may often occur when the closed-loop response of the system needs to be altered in order to meet the performance requirements.

The augmented desired trajectory is obtained through:

$$y_m(k+1) = y_d(k+1) - 0.01(y_d(k) - y(k)) - 0.001(y_d(k-1) - y(k-1)) \tag{21}$$

NNI has five neurons in the hidden layer with three plant and two input time delays. Learning rate of the identifier is selected to be $\alpha = 0.002$, and momentum term equals $\gamma = 0.01$. The NNC has eight neurons in the hidden layer with two time delays, respectively for plat output, plant input and reference input (a 6-8-1 network). Learning rate of the controller is found from $\alpha_c = 3 \times 10^{-4} \exp(0.1 \mid \epsilon \mid)$ and the momentum term is set to $\gamma_c = 0.01$.

Reference input is produced as:

$$r(t) = \sin(t) \tag{22}$$

Sampling times for identifier and controller are chosen equal to 0.001 sec and 0.0001 sec. Fig. 9 represents the response of the controlled plant and the control signal. One may see some spikes in the times response of the system. At the beginning, the spikes happen as a result of sinusoidal input that is applied for few seconds to identify the plant (see Remark 1). The spike at $t = 8$ sec is due to the change of the reference model.

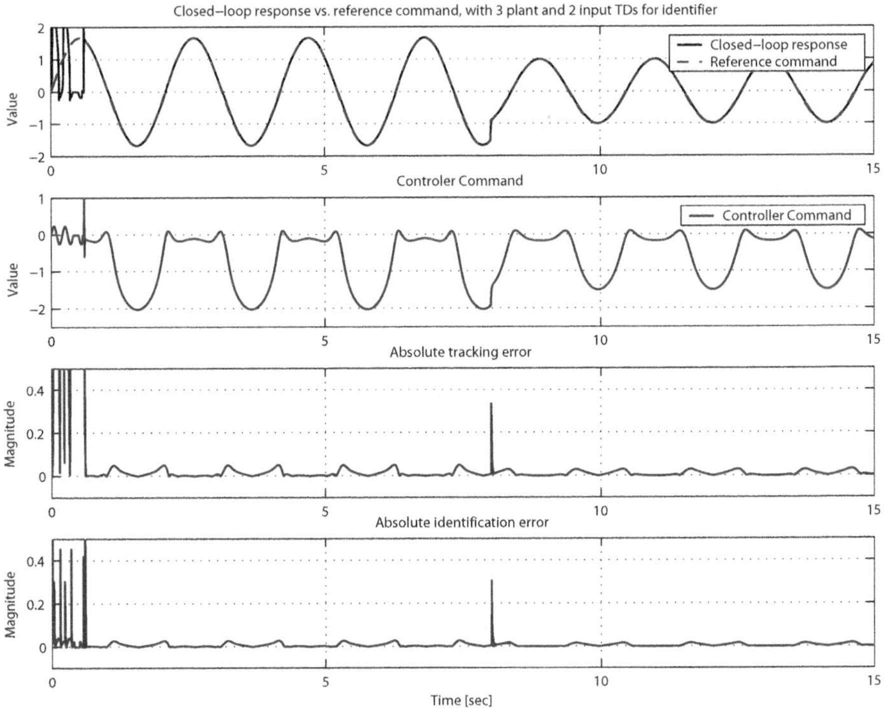

Fig. 9 Time response of the controlled plant in Example 1

5.2 *Example 2: A Non-linear Plant Subjected to Uncertainty*

In this example, the robustness of the proposed control method to plant uncertainty and perturbations is investigated. The performance of a control system in the face of plant uncertainty is an important issue. No mathematical system can exactly model a physical system and as a result it is necessary to be aware of how modeling errors due to the plant uncertainties affect the performance of the control system. Typical sources of uncertainty include unmodeled (high frequency) dynamics, neglected non-linearities, and plant parameter (dynamic) perturbations. If the control system performs well for these types of variations in the system dynamics and the stability of the closed-loop system is maintained, then the scheme is said to be robust.

To demonstrate the robustness of the proposed control scheme, we consider the following plant (39):

$$y(k+1) = f[y(k), y(k-1), y(k-2), u(k), u(k-1)] + \Delta \tag{23}$$

where the function f is given below:

$$f(x_1, x_2, x_3, x_4, x_5) = \frac{x_1 x_2 x_3 x_5 (x_3 - 1) + x_4}{1 + x_2^2 + x_3^2} \tag{24}$$

while the perturbation model Δ used is a random number generated between -0.3 and 0.3 with sampling time of 3 sec. The reference signal is generated by:

$$r(t) = 0.5(0.5\sin(t) + \cos(0.2t)) \tag{25}$$

while the reference models signal is generated by:

$$y_d(t) = \sin(r(t)) \tag{26}$$

The augmentation scheme is given as:

$$y_m(k+1) = y_d(k+1) - 0.1(y_d(k) - y(k)) - 0.01(y_d(k-1) - y(k-1)) \tag{27}$$

The NNI is a 5-5-1 network with three plant and two input time delays with learning rate and momentum term, respectively, equal to $\alpha = 0.002$ and $\gamma = 0.01$; the NNC is a 8-6-1 network (two delays for system output, four delays for controller command and two delays for reference signal) with $\alpha_c = 5 \times 10^{-3}$ and $\gamma_c = 0.01$. Fig. 10 shows both controlled plant's response and controller's command. It is clear that the reference signal is tracked and the disturbance is rejected while influencing the controller's command. The spikes that appears in the identification and tracking error could happen due to the external disturbances applied to the system.

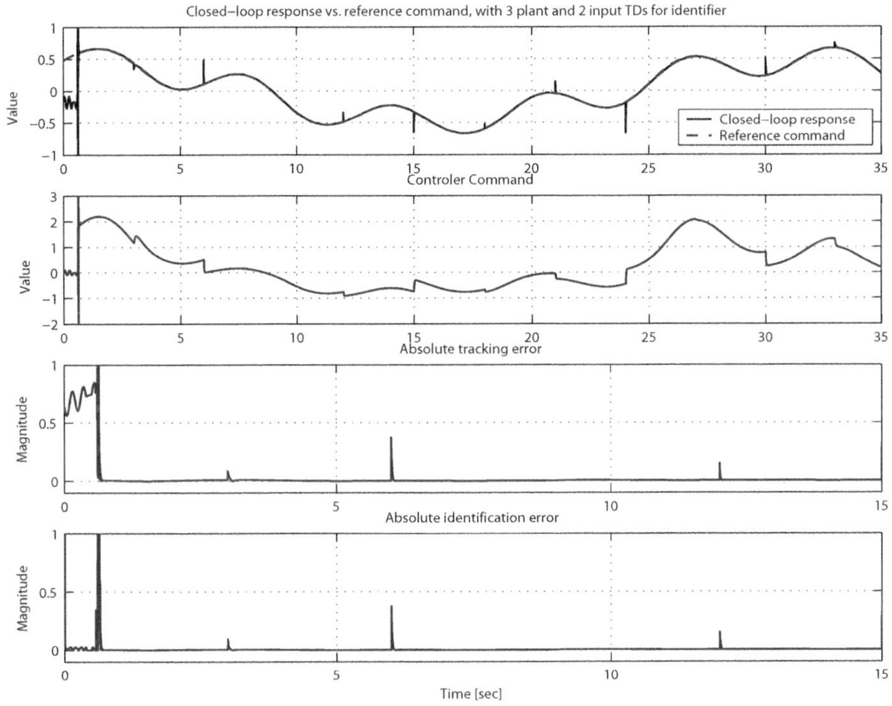

Fig. 10 Time response of the controlled plant in Example 2

6 Stick-Slip Friction Compensation Using the Introduced Neuro-Control Algorithm

6.1 Problem Statement

The system under investigation is a DC-motor driven mechanical system with friction. Consider that position and velocity measurements are taken with the period of discretization $T_{0s} = 0.1$ sec; the system's gain (k_o) and the system's mass (m) are 8 and $1kg$, respectively; the load disturbance depends on the position and the velocity:

$$d(t) = d_1 y(t) + d_2 v(t)$$

where $d_1 = 0.25$ and $d_2 = -0.7$.

The discrete-time model of the 1-DOF mass mechanical system with the friction subjected to load disturbance, given above, is (13), (15):

$$
\begin{aligned}
x_1(k+1) =& x_2(k) \\
x_2(k+1) =& -0.025x_1(k) - 0.3x_2(k) + 0.8u(k) \\
& -0.1F_f(\bullet) \\
v(t) =& x_2(k) - x_1(k) \\
y(k) =& 0.1x_1(k)
\end{aligned}
\tag{28}
$$

where $x_1(k)$ and $x_2(k)$ are the system's states, $v(k)$ is the system's velocity, $y(k)$ is the system's position, $u(k)$ is the control input and $F_f(\bullet)$ is an unknown friction force modelled as given in (8), (9). The meaning of the stick-slip friction parameters and their numerical values in the simulations are given in Table 1. For illustration an example of the slipping friction force plot with $\bar{\alpha} = 0$ is shown in Fig. 11.

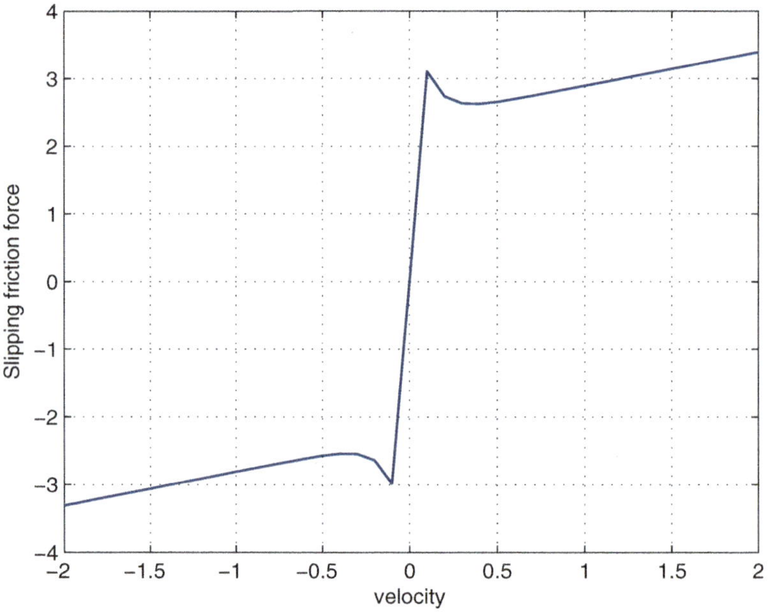

Fig. 11 The graph of the slipping friction force

6.2 Simulation Results

In this section, we present simulation studied to illustrate the performance of the developed adaptive control scheme for stick-slip friction compensation (28). The neuro-control algorithm does not need any special setup or change; it only needs assignment of the two **A** and **B** matrices. Due to instability and unexpected changes in the dynamics of the system, relatively large values

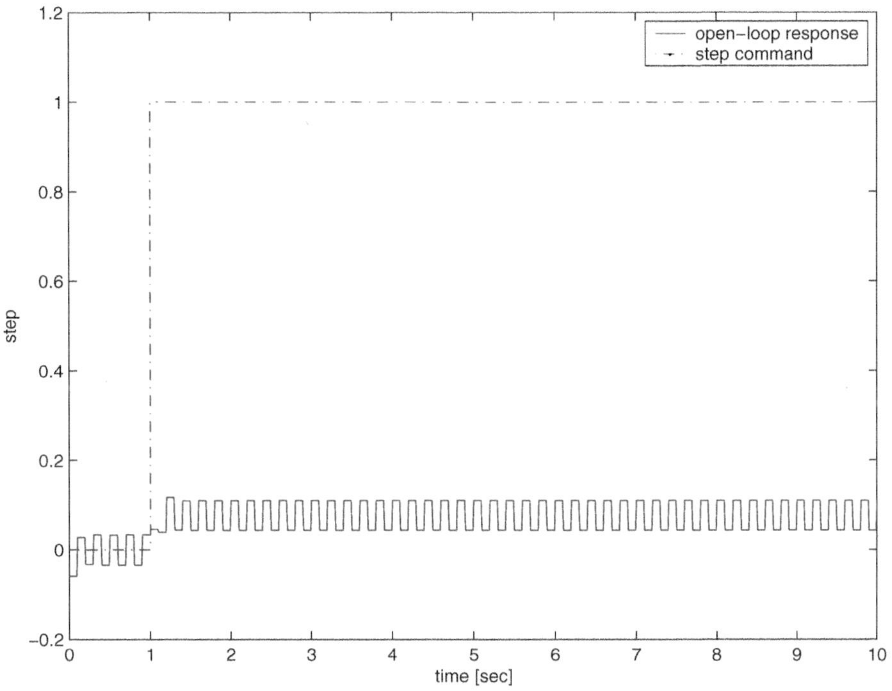

Fig. 12 Step response of the open-loop system

Table 1 Simulation parameters for stick-slip friction model

Symbol	Explanation	Quantity ($0 \leq t < 15$)	Quantity ($t \geq 15$)
$\bar{\alpha}$	Critical stick to slip velocity (m/s)	0.001	0.001
F_s^+	Positive limit on the static friction force (N)	4.2	5.4
F_s^-	Negative limit on the static friction force (N)	-4.0	-5.0
ΔF^+	Positive drop from the static to the kinetic force level (N)	1.8	2.5
ΔF^-	Negative drop from the static to the kinetic force level (N)	-1.7	-2.0
$\dot{x}_{cr}^+$	Positive critical Stribeck velocity (m/s)	0.1	0.2
$\dot{x}_{cr}^-$	Negative critical Stribeck velocity (m/s)	0.1	0.2
b^+	Positive viscous friction coefficient (N.s/m)	0.5	0.4
b^-	Negative viscous friction coefficient (N.s/m)	0.5	0.4

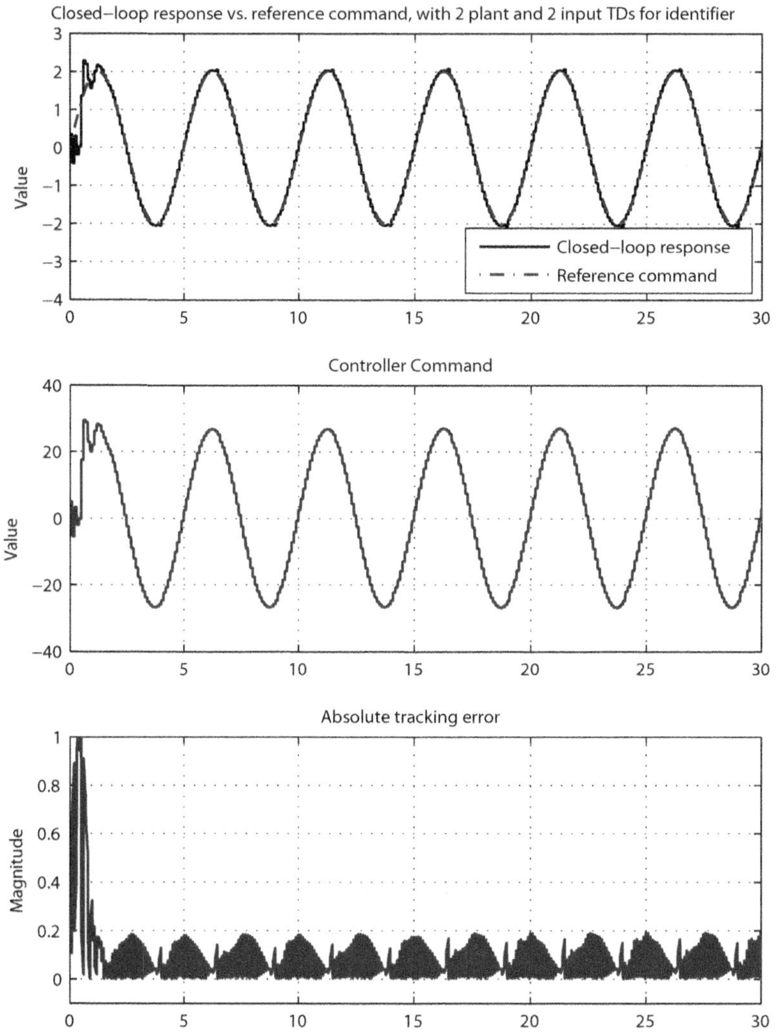

Fig. 13 Tracking performance of the proposed control scheme to a sinusoidal input (upper diagram); Control effort (middle diagram); Tracking error (lower diagram)

for the two **A** and **B** matrices are chosen: **A** = 0.8, and **B** = 0.1. So, the augmentation scheme becomes:

$$y_m(k+1) = y_d(k+1) - 0.8(y_d(k) - y(k)) - 0.1(y_d(k-1) - y(k-1)) \tag{29}$$

The NNI has a structure of 4-4-1, with 2 plant and 2 input time delays with learning rate and momentum term equal to and $\alpha = 0.002$ and $\gamma = 0.1$,

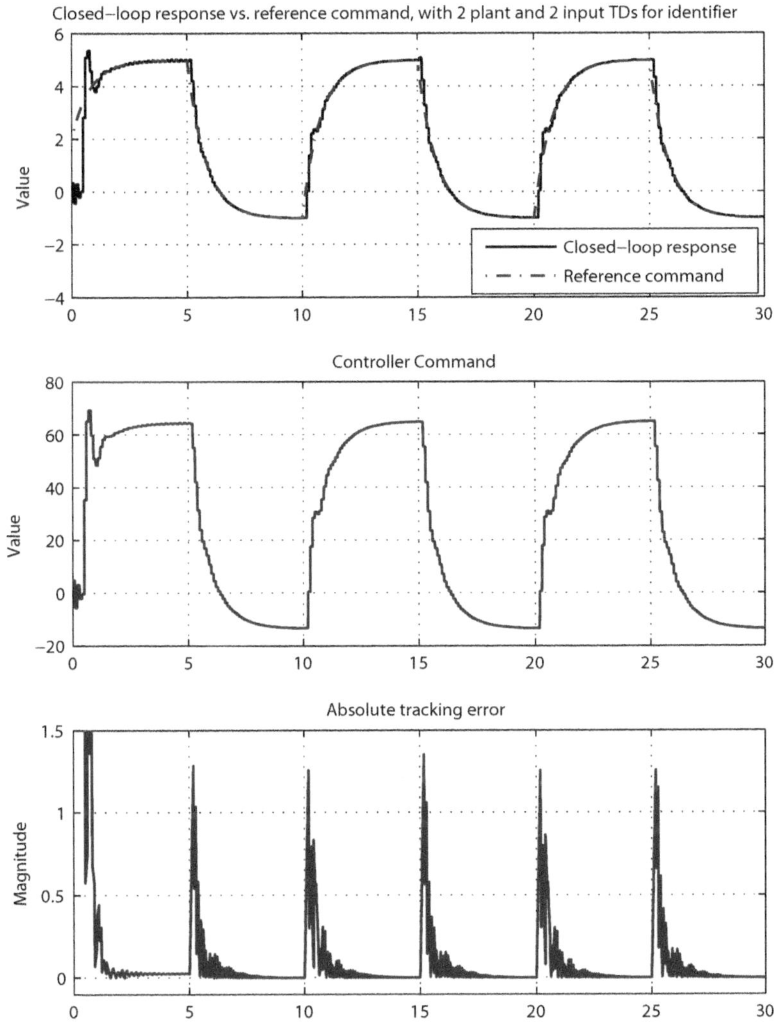

Fig. 14 Tracking performance of the proposed control scheme to step reference command (upper diagram); Control effort (middle diagram); Tracking error (lower diagram)

respectively. The NNC is of 6-5-1 structure, with the learning rate set as and the momentum term $\alpha_c = 8e - 2$ and $\gamma_c = 0.05$. Sampling time for identifier is 0.01 sec and for controller is equal to 0.1 sec.

The time response of the open-loop system (28) to a step command of the input signal is shown in Fig. 12. It is clear in Fig. 12 that the open-loop system has a very poor tracking performance; it shows an oscillatory response due to load disturbance as well. To study the performance of the proposed control scheme, first tracking of a sinusoidal input is considered. The input signal and the closed-loop system's response are shown in Fig. 13. Note that the parameters of the friction force are changed at time $t = 15$ sec as given in Table 1, to show the adaptability and the robustness of the proposed control scheme. From Fig. 13, it can be seen that the proposed controller has a very good performance in tracking of the commanded signal, which is due to the controller's adaptability and robustness.

The performance of the proposed control scheme is also investigated for tracking of a series of step input commands. The linear reference model is used with the transfer function:

$$[TF]_{\text{reference}} = \frac{1}{0.8s + 1} \tag{30}$$

The reference position trajectory, the closed-loop system's response, the controller command and the tracking error are shown in Fig. 14. It should be mentioned that whenever the velocity becomes zero, the control input becomes smaller than the friction force, so the mass sticks. However, such a situation is avoided by using the proposed controller (the tracking error converges to zero). The proposed scheme is robust to the variations of the system and/or stick-slip friction characteristics and the bounded external disturbance (implemented in the systems system's dynamics).

7 Conclusions

In this paper, a general purpose neural-network-based adaptive controller is introduced. The developed algorithm essentially realizes an indirect-adaptive control scheme with the use of a neural network. The adopted neural network is implemented for identification of an unknown process and a neuro-controller, which uses the data provided by the identifier for decision making. The stability of the control scheme is proved using Lyapunov's direct method. The learning algorithm used in the proposed scheme to adjust the parameters of the networks is momentum back-propagation (MBP). This scheme has the following properties: (1) it does not require any pre-training data and robustifying terms; this in turn helps the proposed neural-control method to compensate the abrupt changes in the system, which may happen due to environmental and unforeseen conditions, as well as the measurement noise; (2) it needs a small number of tuning parameters requiring a few trial and error to be properly selected. This may be considered as the disadvantage of the proposed scheme. Another disadvantage of the proposed scheme is that it is not suitable for systems with time delays, which is due to the fact that the output of the plant is not influenced by the controller's command at the present time. Therefore, the back-propagation through the plant that is used to update the weights of the controller fails to propagate the influence of the input signal imposed on the plant to its output. This indeed will result in the overall system destabilization.

The simulation results help us better evaluate the merits of the proposed neuro-control scheme. It has been demonstrated that the outputs of the unknown plants can be controlled by closely tracking the reference signals, despite the fact that the structure of the plant might vary and/or be subject to noise. Furthermore, it is shown that the control scheme is capable of effectively dealing with stable reference models with time varying dynamics.

As for practical purposes, the introduced control algorithm could be employed for compensation of stick-slip friction in electro-mechanical systems. It is shown that the algorithm can effectively eliminate the impact of the friction on the system's performance without any major modifications made to the algorithm.

References

1. Dupont, P.E.: Avoiding stick-slip through PD control. IEEE Trans. Automatic Contr. 39, 1094–1097 (1994)
2. Leonard, N.E., Krishnaprasad, P.: Adaptive friction compensation for bi-directional low velocity position tracking. In: Proc. 31st CDC, Tucson, AZ, pp. 267–273 (1992)
3. Friedland, B., Park, Y.J.: On adaptive friction compensation. IEEE Trans. Automatic Contr. 40, 419–425 (1992)
4. Li, W., Cheng, X.: Adaptive high-precision control of positioning tables-Theory and experiments. IEEE Trans. Contr. Syst. Technol. 2, 265–270 (1994)
5. Tan, K.K., Lim, S.Y., Lee, T.H., Huang, S.N.: Adaptive control of DC permanent magnet linear motor for ultra-precision applications. In: Int. Conf. Mechatronic Tech., Taiwan, R.O.C., pp. 243–246 (1998)
6. De Wit, C.C., Lischinsky, P.: Adaptive friction compensation with partially known dynamic friction model. Int. J. Adapt. Contr. Sig. Process. 11(1), 65–80 (1997)
7. Otten, G., de Vries, T.J.A., van Amerongen, J., Rankers, A.M., Gaal, E.W.: Linear motor motion control using a learning forward controller. IEEE/ASME Trans. Mechatronics 2(3), 179–187 (1997)
8. Southward, S.C., Radcliffe, C.J., MacCluer, C.R.: Robust nonlinear stick-slip friction compensation. ASME J. Dyn. Syst., Meas., and Contr. 113, 639–645 (1991)
9. Lee, S.-W., Kim, J.-H.: Robust adaptive stick-slip friction compensation. IEEE Trans. Industrial Elec. 42(5) (1995)
10. De Wit, C.C., Olsson, H., Astrom, K., Lischinsky, P.: A new model for control of systems with friction. IEEE Trans. Automatic Contr. 40, 419–425 (1995)
11. Panteley, E., Ortega, R., Gafvert, M.: An adaptive friction compensator for global tracking in robot manipulators. Syst. Contr. Lett. 33, 307–313 (1998)
12. Huang, S.N., Tan, K.K., Lee, T.H.: Adaptive friction compensation using neural network approximations. IEEE Trans. Syst., Man and Cyber.-C 30(4), 551–557 (2000)

13. Baruch, I., Beltran, R.L., Olivares, J.-L., Garrido, R.: A Fuzzy-Neural Multi-Model for Mechanical Systems Identification and Control. In: Monroy, R., Arroyo-Figueroa, G., Sucar, L.E., Sossa, H. (eds.) MICAI 2004. LNCS (LNAI), vol. 2972, pp. 774–783. Springer, Heidelberg (2004)
14. Suraneni, S., Kar, I.N., Bhatt, R.K.P.: Adaptive stick-slip friction compensation using dynamic fuzzy logic system. In: Proc. TENCON 2003, the conference on convergent Technologies for Asia-Pacific Region, vol. 4, pp. 1470–1474 (2003)
15. Baruch, I., Garrido, R., Mitev, A., Nenkova, B.: A neural network approach for stick-slip model identification. In: Proc. of the 5th Int. Conf. on Engineering Applications of Neural Networks, EANN 1999, Warsaw, Poland, pp. 183–188 (1999)
16. Cohen, M.A., Grossberg, S.: Absolute stability of global pattern formation and parallel memory storage by competitive neural networks. IEEE Trans. on Systems, Man, and Cybernetics 13(5), 815–825 (1993)
17. Hopfield, J.J.: Neural networks and physical systems with emergent collective computational abilities. Proc. of the National Academy of Sciences, USA 79, 2554–2558 (1982)
18. Kosko, B.: Structural stability of unsupervised learning in feedback neural networks. IEEE Trans. on Automatic Control 36(7), 785–792 (1991)
19. Guez, A., Protopopsecu, V., Barhen, J.: On the stability, storage capacity and design of non-linear continuos neural networks. IEEE Trans. on Systems, Man, and Cybernetics 18(1), 80–87 (1988)
20. Kelly, D.G.: Stability in contractive non-linear neural networks. IEEE Trans. on Biomedical Eng. 37(3), 231–242 (1990)
21. Vidyasagar, M.: Location and stability of the high gain equilibria of non-linear neural networks. IEEE Trans. on Neural Networks 4(4), 660–672 (1993)
22. Jim, L., Nikiforuk, P.N., Gupta, M.M.: Absolute stability conditions for discrete-time recurrent neural networks. IEEE Trans. on Neural Networks 5(6), 954–964 (1994)
23. Tzirkel-Hancock, E., Fallside, F.: A stability based neural network control method for a class of non-linear systems. In: Proc. Inter. Joint Conf. on Neural Networks, vol. 2, pp. 1047–1052 (1991)
24. Lewis, F.L., Yesildirek, A., Liu, K.: Neural net robot controller with guaranteed stability. In: Proc. 3rd Inter. Conf. on Indus. Fuzz. Cont., pp. 103–108 (1993)
25. Lewis, F.L., Yesildirek, A., Liu, K.: Multilayer neural net robot controller with guaranteed tracking performance. IEEE Trans. on Neural Networks 7(2), 388–399 (1996)
26. Kwan, C., Lewis, F.L., Dawson, D.M.: Robust neural-network control of rigid-link electrically driven robots. IEEE Trans. on Neural Networks 9(4), 581–588 (1998)
27. Kuntanapreeda, S., Fullmer, R.R.: A training rule which guaranteed finite region stability for a class of closed-loop neural network control system. IEEE Trans. on Neural Networks 7(3), 629–642 (1996)
28. Park, S., Park, C.H.: Comments on a training rule which guarantees finite-region stability for a class of closed-loop neural-network control systems. IEEE Trans. on Neural Networks 8(5), 1217–1218 (1997)
29. Sadegh, N.: A perceptron network for functional identification and control of non-linear systems. IEEE Trans. on Neural Networks 4(6), 982–988 (1993)

30. Polycarpou, M.M.: Stable adaptive neural control scheme for non-linear systems. IEEE Trans. Auto. Control 41(3), 447–451 (1996)
31. Chen, F.-C., Khalil, H.K.: Adaptive control of non-linear systems using neural networks. Int. J. Control 55, 1299–1317 (1992)
32. Chen, F.-C., Khalil, H.K.: Adaptive control of a class of Non-linear Discrete-Time Systems Using Neural Networks. IEEE Trans. Auto. Control 40(5), 791–801 (1995)
33. Chen, F.-C., Liu, C.-C.: Adaptively controlling non-linear continuous-time systems using multilayer neural networks. IEEE Trans. Auto. Control 39(6), 1306–1310 (1994)
34. Fabri, S., Kadirkamanathan, V.: Dynamic structure neural networks for stable adaptive control of non-linear systems. IEEE Trans. on Neural Networks 7(5), 1151–1167 (1996)
35. Sun, F., Sun, Z., Woo, P.-Y.: Stable neural-network-based adaptive control for sampled-data non-linear systems. IEEE Trans. on Neural Networks 9(5), 956–968 (1998)
36. Jagannathan, S., Lewis, F.L.: Multilayer discrete-time neural-net controller with guaranteed performance. IEEE Trans. on Neural Networks 7(1), 107–130 (1996)
37. Jagannathan, S., Lewis, F.L., Pasravanu, O.: Discrete-time model reference adaptive control of nonlinear dynamical systems using neural networks. Int. J. Contr. 64(2), 217–239 (1996)
38. Suykens, J.A.K., Vandewalle, J.: Global asymptotic stability for multilayer recurrent neural networks with application to modelling and control. In: Proc. Inter. Conf. on Neural Network., vol. 2, pp. 1065–1069 (1995)
39. Narendra, K.S., Parthasarathy, K.: Identification and control of dynamical systems using neural networks. IEEE Trans. on Neural Net. 1(1), 4–27 (1990)
40. Mehrabian, A.R., Menhaj, M.B.: A real-time neuro-adaptive controller with guaranteed stability. Applied Soft Comput. J. 8(1), 530–542 (2008)
41. Hagan, M.T., Menhaj, M.: Training feedforward networks with the Marquardt algorithm. IEEE Trans. on Neural Net. 5(6), 989–993 (1994)
42. Sinha, N.K., Rao, G.P.: Identification of Continuous-time Systems: Methodology and Computer Implementation. Kluwer Academic, Dordrecht (1991)
43. Hornik, K.M., Stinchcombe, M., White, H.: Multilayer feedforward networks are universal approximators. Neural Net. 2(5), 359–366 (1989)
44. Hagan, M., Demuth, H., Beale, M.: Neural Network Design. PWS Publishing, Boston (1996)

Modeling of Crude Oil Blending via Discrete-Time Neural Networks

Xiaoou Li and Wen Yu

1 Introduction

Crude oil blending is an attractive solution for those refiners who have the ability to blend different crude types to provide a consistent and optimal feedstock to refinery operations. Optimal crude purchasing is an effective method to improve refinery profits. In general the blending rule is nonlinear, it can be regarded as a linear mixing rule adding a nonlinear term. Crude oil blending is an optimization operation based upon real-time analyzers and process knowledge (6). A mathematical model for crude oil blending is needed to address uncertainties in blending operation, real-time optimization (RTO) has been proposed (20). The main drawback of RTO is that it cannot provide optimal set-points from large amounts of history data.

The exact mathematical model for crude oil blending is too complex to be handled analytically. Many attempts were made to introduce simplified models to construct "model-based" controller (9). A common method to approximate the blending operation is to use a linear (ideal) model (20) or to regard blending operation to have a sufficient small nonlinear uncertainty (1).

A neuro modeling approach uses the nice features of neural networks, but the lack of a mathematical model for the plant makes it hard to obtain theoretical results on stable learning. It is very important to assure the stability of neuro modeling in theory before we use it in some real applications. Lyapunov approach can be used directly to obtain robust training algorithms for continuous-time (23)(24) and discrete-time (13)(18) neural networks. It is well known that normal modeling algorithms are stable for ideal plants (11). In the presence of disturbances or unmodeled dynamics, these adaptive procedures can go to instability easily. Generally, some modifications to the normal gradient algorithm or backpropagation should be applied, such that the learning process is stable. For example, in (13) some hard restrictions were added in the learning law, in (22) the dynamic backpropagation was modified with NLq stability constraints. Another generalized method is

J. Schumann, Y. Liu (Eds.): Appl. of Neural Networks in High Assur. Sys., SCI 268, pp. 205–220.
springerlink.com

to use robust modification techniques of robust adaptive control (11). (15) applied $\sigma-$modification, and (12) used a modified $\delta-$rule.

In this paper, we propose a novel learning algorithm for discrete-time feedforward neural network. By combining Lyapunov and dead-zone techniques, we analyze the stability of modeling error and the parameters. This learning law can guarantee that both modeling error and weights are bounded. The neuro modeling approach is successfully used to model crude oil blending via real data.

2 Crude Oil Blending

Crude oils are often blended to increase the sale price or process ability of a lower grade crude oil by blending it with a higher grade, higher price crude. The objective is to produce blended crude oil to a target specification at the lowest cost using the minimum higher cost crude oil. The crude oil feedstocks used for blending often vary in quality and for this reason crude oil blenders normally use viscosity or density trim control systems. API (American Petroleum Institute) Gravity is the most used indication of density of crude oil. The lower the API Gravity, the heavier the compound. When the blender is started the required flow rate and component ratio is set by the control system based on the ratio in the recipe. A density or viscosity analyzer, installed at a homogeneous point in the blender header, generates a control signal, which is used to continually optimize the blended product by adjusting the component ratio. This ensures that the blended product remains as specified at all times during the batch. So normal modeling for crude oil blending is on-line. In this paper we will discuss an off-line modeling method.

We discuss a typical crude oil blending process in PEMEX (Mexican Petroleum Company), which is called Terminal Marítima de Dos Bocas Tabasco (TMDB). The flow diagram is shown in Fig. 1-(a). It has three blenders (M_1, M_2, and M_3), one dehydration equipment and one tank. We use Fig. 1-(b) to describe the static process of the crude oil blending, q_i is flow rate, p_i is the property of ith feed stock, it can be API Gravity. There are four feed-stocks, $L_3\,(q_1, p_1)$, Puerto Ceiba (q_2, p_2), Maya (q_3, p_3) and El Golpe (q_4, p_4). The blended product for national use (q_n, p_n) needs 2 blenders, the product for export (q_f, p_f) needs 3 blenders and is stored in a tank.

For each blender, static properties can be analyzed by thermodynamic theory. If the partial molar volume of a component in a solution is nearly the same as the molar volume of the pure component, the molar volume is simply the average of the molar volumes of the pure components at the same temperature and pressure. The thermodynamic property is ideally

$$p_a = \sum_{i=1}^{2} p_i x_i, \quad q_a = \sum_{i=1}^{2} q_i, \quad x_i = \frac{q_i}{q_a}$$

where x_i, q_i and p_i are the volume fraction, flow rate and API Gravity of ith feed-stock, p_a and q_a are the flow rate and the API Gravity of the blended product of Blender 1 (M_1). Unfortunately, this equation is correct only under the ideal condition, in order to make it universally valid a correction term Δ is added

$$p_a = \sum_{i=1}^{2} p_i x_i + \Delta \tag{1}$$

where Δ is called the property change of mixing. Several approaches can approximate Δ, for example

- Interaction model (20)(1)
$$\Delta = \alpha x_1 x_2 \tag{2}$$
  where α is the interaction coefficient between the two components
- Zahed model (27)
$$\Delta = \sum_{i=1}^{2} M_i (x_i p_i)^k \tag{3}$$
  where M_i and k are constants.

All of above models are only suitable under some special conditions and the parameters of these models should be determined by experimental data.

Since all of p_i and q_i in Fig. 1-(b) are available, we can model each blender with input/output data, then connect them together, we call this method as

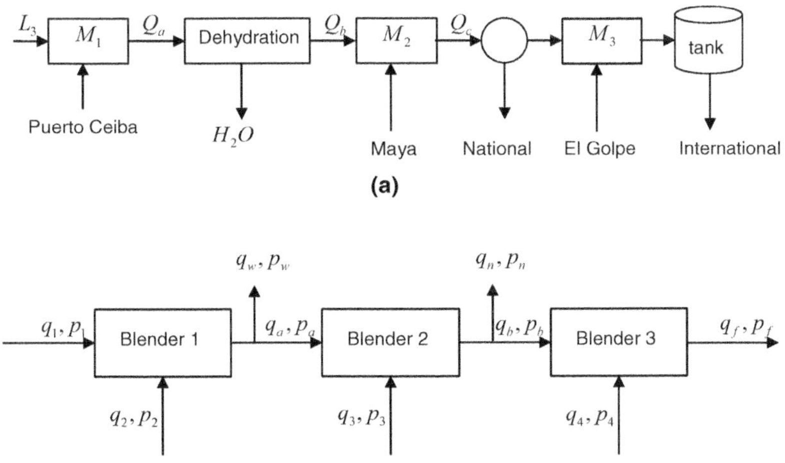

Fig. 1 TMDB crude oil blending process

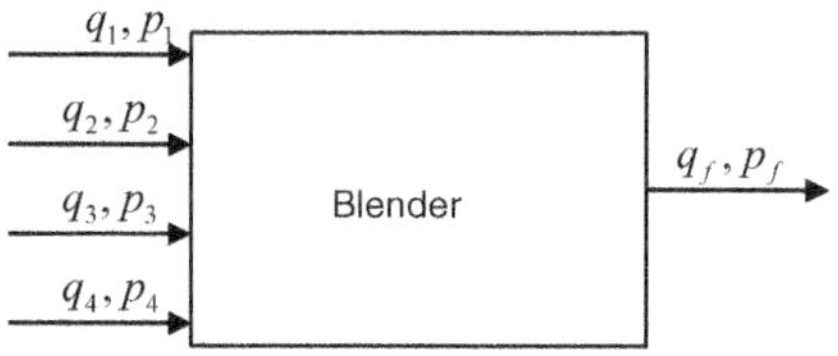

Fig. 2 Integrated model

distribute model. If the mixing rule is given by a interaction model as in (2), the total blending is

$$\begin{aligned}
p_f &= \tfrac{1}{q_f}\left(p_4 q_4 + p_b q_b\right) + \alpha_3 x_4 x_b \\
&= \tfrac{1}{q_f} p_4 q_4 + \tfrac{1}{q_f}[p_3 x_3 + \left(\tfrac{q_a}{q_a}\left(p_2 q_2 + p_1 q_1\right) + \alpha_1 x_1 x_2\right) \\
&(1 - x_3) + \alpha_2 x_3 (1 - x_3)](1 - x_4) + \alpha_3 x_4 (1 - x_4) \\
&= \tfrac{1}{q_f}\left(p_1 q_1 + p_2 q_2 + p_3 q_3 + p_4 q_4\right) \\
&+ \tfrac{1}{q_f}[x_3 \alpha_2 - p_1 x_1 x_3 - p_1 x_1 x_4 \\
&- p_2 x_2 x_3 - p_2 x_2 x_4 - p_3 x_3 x_4 \\
&+ x_1 x_2 \alpha_1 - x_3 x_4 \alpha_2 + p_1 x_1 x_3 x_4 + p_2 x_2 x_3 x_4 \\
&- x_1 x_2 x_3 \alpha_1 - x_1 x_2 x_4 \alpha_1 + x_1 x_2 x_3 x_4 \alpha_1 - x_3^2 \alpha_2 \\
&+ x_3^2 x_4 \alpha_2 + x_4 \alpha_3 - x_4^2 \alpha_3]
\end{aligned} \tag{4}$$

where $q_f = q_1 + q_2 + q_3 + q_4 - q_w - q_n$, α_i is mixing rule coefficient for i-th blender.

We can also regard the model as a multiple components blending process as in Fig. 2, we call it as *integrated model.* The model can be expressed as

$$p_f = \sum_{i=1}^{4} p_i x_i + \Delta \tag{5}$$

If the mixing rule is given by a interaction model as in (2)

$$\Delta = \sum_{i=1}^{4} \sum_{k=i+1}^{4} \alpha_{i,k} x_i x_k \tag{6}$$

3 Modeling of Crude Oil Blending via Discrete-Time Neural Networks

The mathematical models discussed in Sect. 2 work only in some special conditions. In real applications we have only input/output data. A neural network can be applied to identify crude oil blending. Static neural networks can be used to identify the nonlinear parts Δ of the distribute model (1) or the integrated model (5), it can also identify the whole blender (linear and nonlinear). This section will present a new stable learning algorithm for static neuro modeling.

The mixing property can be written in following form
$p_f(k) = \Phi[u_1(k), \cdots, u_8(k)]$, or

$$y(k) = \Phi[X(k)] \tag{7}$$

where $X(k) = [u_1(k), \cdots, u_8(k)]^T$, $y(k)$ is the blended API Gravity value at time k, $y(k) = p_f(k)$, $\Phi(\cdot)$ is an unknown nonlinear function representing the blending operation, $u_i(k)$ are measurable scalar inputs, API Gravity and flow rates, for example $u_1(k) = \frac{q_1}{q_o}$, $u_2(k) = p_1$, $u_7(k) = \frac{q_4}{q_o}$, $u_8(k) = p_4$. We consider a multilayer neural network (or multilayer perceptrons) to model the blending properties as in (7)

$$\widehat{y}(k) = V_k \phi[W_k X(k)] \tag{8}$$

where the scalar output $\widehat{y}(k)$ and vector input $X(k) \in R^{n\times 1}$, the weights in output layer are $V_k \in R^{1\times m}$, the weights in hidden layer are $W_k \in R^{m\times n}$, ϕ is $m-$dimension vector function. The typical presentation of the element $\phi_i(.)$ is a sigmoid function. The identified blending system (7) can be represented as

$$y(k) = V^* \phi[W^* X(k)] - \mu(k)$$

where V^* and W^* are set of unknown weights, which may minimize the modeling error $\mu(k)$. The nonlinear plant (7) can be also expressed as

$$y(k) = V^0 \phi[W^* X(k)] - \delta(k) \tag{9}$$

where V^0 is an known matrix chosen by users. In general, $\|\delta(k)\| \geq \|\mu(k)\|$. Using a Taylor series around the point of $W_k X(k)$, the modeling error can be represented as

$$\begin{aligned} e(k) &= \widehat{y}(k) - y(k) \\ &= V_k \phi[W_k X(k)] - V^0 \phi[W^* X(k)] + \delta(k) \\ &= V_k \phi[W_k X(k)] - V^0 \phi[W_k X(k)] \\ &\quad + V^0 \phi[W_k X(k)] - V^0 \phi[W^* X(k)] + \delta(k) \\ &= \widetilde{V}_k \phi[W_k X(k)] + V^0 \phi' \widetilde{W}_k X(k) + \zeta(k) \end{aligned} \tag{10}$$

where ϕ' is the derivative of nonlinear activation function $\phi(\cdot)$ at the point of $W_k X(k)$, $\widetilde{W}_k = W_k - W^*$, $\widetilde{V}_k = V_k - V^0$, $\zeta(k) = V^0 \varepsilon(k) + \delta(k)$, here $\varepsilon(k)$ is second order approximation error of the Taylor series.

In this paper we are only interested in open-loop modeling. So we can assume that the plant (7) is bounded-input and bounded-output stable, *i.e.*, $y(k)$ and $u(k)$ in (7) are bounded.

Since $X(k) = [u(k), u(k-1), u(k-2), \cdots]^T$, $X(k)$ is bounded. By the boundedness of the sigmoid function ϕ, we assume that $\delta(k)$ in (9) is bounded, also $\varepsilon(k)$ is bounded. So $\zeta(k)$ in (10) is bounded. The following theorem gives a new robust learning algorithm and stable analysis for the neural modeling.

Theorem 1. *If we use the multilayer neural network (8) to model the crude oil blending (7), the following dead-zone backpropagation-like algorithm*

$$\begin{array}{l} W_{k+1} = W_k - \eta_k e(k) \phi' V^{0T} X^T(k) \\ V_{k+1} = V_k - \eta_k e(k) \phi^T \end{array} \tag{11}$$

where $\eta_k = \dfrac{s_k}{1 + \|\phi' V^{0T} X^T(k)\|^2 + \|\phi\|^2}$, $s_k = \begin{cases} \eta \; e(k)^2 \geq \frac{\eta}{\pi}\overline{\zeta} \\ 0 \; e(k)^2 < \frac{\eta}{\pi}\overline{\zeta} \end{cases}$, $0 < \eta \leq 1$, $\pi = \dfrac{\eta}{(1+\kappa)^2} > 0$, $\overline{\zeta} = \max_k \left[\zeta^2(k)\right]$, $\kappa = \max_k \left(\|\phi' V^{0T} X^T(k)\|^2 + \|\phi\|^2 \right)$.
This updating law can make the modeling error $e(k)$ *and the weights of neural networks bounded*

$$\|e(k)\| \in L_\infty, \quad W_k \in L_\infty, \quad V_k \in L_\infty \tag{12}$$

Also the average of the modeling error satisfies

$$J = \limsup_{T \to \infty} \frac{1}{T} \sum_{k=1}^{T} e^2(k) \leq \frac{\eta}{\pi} \overline{\zeta} \tag{13}$$

where $\pi = \dfrac{\eta}{1+\kappa}\left[1 - \dfrac{\eta\kappa}{1+\kappa}\right] > 0$, $\kappa = \max_k \left(\|\phi' V^{0T} X^T(k)\|^2 + \|\phi\|^2 \right)$, $\overline{\zeta} = \max_k \left[\zeta^2(k)\right]$.

Proof. If $e(k)^2 \geq \frac{\eta}{\pi}\overline{\zeta}$, the updating law is (11) with

$\eta_k = \dfrac{\eta}{1 + \|\phi' V^{0T} X^T(k)\|^2 + \|\phi\|^2}$. We selected a positive defined matrix L_k as

$$L_k = \left\|\widetilde{W}_k\right\|^2 + \left\|\widetilde{V}_k\right\|^2 \tag{14}$$

where $\widetilde{W}_k = W_k - W^*$, $\widetilde{V}_k = V_k - V^0$, $\left\|\widetilde{W}_k\right\|^2 = \sum_{i=1}^n \widetilde{w}_k{}^2 = tr\left\{\widetilde{W}_k^T \widetilde{W}_k\right\}$. From the updating law (11), we have

$$\begin{array}{l} \widetilde{W}_{k+1} = \widetilde{W}_k - \eta_k e(k) \phi' V^{0T} X^T(k) \\ \widetilde{V}_{k+1} - \widetilde{V}_k - \eta_k e(k) \phi^T \end{array}$$

Since ϕ' is a diagonal matrix, and by using (10) we have

$$\begin{array}{l}\Delta L_k = \left\| \widetilde{W}_k - \eta_k e(k) \phi' V^{0T} X^T(k) \right\|^2 \\ + \left\| \widetilde{V}_k - \eta_k e(k) \phi^T \right\|^2 - \left\| \widetilde{W}_k \right\|^2 - \left\| \widetilde{V}_k \right\|^2 \\ = \eta_k^2 e^2(k) \left(\left\| \phi' V^{0T} X^T(k) \right\|^2 + \|\phi\|^2 \right) \\ -2\eta_k \|e(k)\| \left\| V^0 \phi' \widetilde{W}_k X(k) + \widetilde{V}_k \phi \right\| \end{array} \tag{15}$$

By (10) we know

$$e(k) = \widetilde{V}_k \phi \left[W_k X(k) \right] + V^0 \phi' \widetilde{W}_k X(k) + \zeta(k)$$

Since $\eta > \eta_k > 0$, the last term in (15) is

$$\begin{array}{l} 2\eta_k \|e(k)\| \left\| V^0 \phi' \widetilde{W}_k X(k) + \widetilde{V}_k \phi \right\| \\ = 2\eta_k \|e(k) \left[e(k) - \zeta(k) \right] \| \\ \geq 2\eta_k e^2(k) - 2\eta_k \|e(k) \zeta(k)\| \\ \geq 2\eta_k e^2(k) - \eta_k e^2(k) - \eta_k \zeta^2(k) \\ \geq \eta_k e^2(k) - \eta \zeta^2(k) \end{array}$$

So

$$\begin{array}{l} \Delta L_k \leq -\eta_k e^2(k) \\ \left[1 - \eta_k \left(\left\| \phi' V^{0T} X^T(k) \right\|^2 + \|\phi\|^2 \right) \right] + \eta \zeta^2(k) \\ \leq -\pi e^2(k) + \eta \zeta^2(k) \end{array} \tag{16}$$

where π is defined in (13). Because $e(k)^2 \geq \frac{\eta}{\pi}\overline{\zeta}$, $\Delta L_k \leq 0$, L_k is bounded, so W_k and V_k are bounded. By (10) $e(k)$ is bounded. If $e(k)^2 < \frac{\eta}{\pi}\overline{\zeta}$, $W_{k+1} = W_k$ and $V_{k+1} = V_k$, so W_k and V_k are bounded, $\|e(k)\|^2 < \frac{\eta}{\pi}\overline{\zeta} < \infty$ is also bounded. For all $e(k)$, (12) is correct.

If $e(k)^2 \geq \frac{\eta}{\pi}\overline{\zeta}$, (16) can be rewritten as

$$\Delta L_k \leq -\pi e^2(k) + \eta \zeta^2(k) \leq -\pi e^2(k) + \eta \overline{\zeta} \tag{17}$$

Summing (17) from 1 up to T, and by using $L_T > 0$ and L_1 is a constant, we obtain

$$\begin{array}{c} L_T - L_1 \leq -\pi \sum_{K=1}^{T} e^2(k) + T\eta\overline{\zeta} \\ \pi \sum_{K=1}^{T} e^2(k) \leq L_1 - L_T + T\eta\overline{\zeta} \leq L_1 + T\eta\overline{\zeta} \end{array}$$

Combing with $e(k)^2 < \frac{\eta}{\pi}\overline{\zeta}$, for all $e(k)$ (13) is established.

Remark 1. V^0 does not effect the stability property of the neuro modeling, but it influences the modeling accuracy, see (13). We design an off-line method to find a better value for V^0. If we let $V^0 = V_0$, the algorithm (11) can make the modeling error convergent, *i.e.*, V_k will make the modeling error smaller than that of V_0. V^0 may be selected by the following steps:

1. Start from any initial value for $V^0 = V_0$, $k = 1$, set window length as T_0
2. Update V_t by the learning law (11), until $k = T_0$.
3. If $\|e(T_0)\| < \|e(1)\|$, let V_T as a new V^0, $V^0 = V_{T_0}$, go to 2 to repeat the modeling process.
4. If $\|e(T_0)\| \geq \|e(1)\|$, stop this off-line modeling, now V_{T_0} is the final value for V^0.

Remark 2. Since we assume neural networks cannot match nonlinear systems exactly, we can not make the parameters (weights) to converge, we would like only to force the output of neural networks to follow the output of the plant, i.e., the modeling error is stable. Although the weights cannot converge to their optimal values, (13) shows that the modeling error will convergence to the ball with radius $\frac{\eta}{\pi}\overline{\zeta}$. Even if the input is persistent exciting, the modeling error $\zeta(k)$ will not make the weights converge to their optimal values. It is possible that the output error is convergent, while the weight errors are very high. The relations of the output error and the weight errors are shown in (10). A simpler case is that weights are parameter-in-linear and the neural networks can match the nonlinear plant exactly

$$\begin{array}{l} \text{plant: } y = W^* \phi[X(k)] \\ \text{neural networks: } \widehat{y} = W_t \phi[X(k)] \\ \text{output error: } (y - \widehat{y}) = (W^* - W_t)\,\phi[X(k)] \end{array}$$

A small output error $(y - \widehat{y})$ does not mean good convergence of the weight error $(W^* - W_t)$ due to $\phi[X(k)]$. Although this algorithm can only be applied to the feedforward neural network with one hidden layer, it can be extended to general form of multilayer perceptions with back-propagation technique.

Remark 3. Noise (or disturbance) is an important issue in the system modeling. There are two types disturbances: external and internal. Internal disturbance can be regarded as unmodeled dynamic $\delta(k)$ in (9). A bounded internal disturbance does not effect the theory results in this paper, but can enlarge the modeling error if the internal disturbance becomes bigger. External disturbance can be regarded as measurement noise, input noise, etc. In the point of structure, input noises are increased feedforward through each layer (2). For example, a noise $\varsigma(k)$ is multiplied by $V_k\phi[W_k\varsigma(k)]$ and arrives at the output. Measurement noise is enlarged due to backpropagation of modeling error (11), therefore the weights of neural networks are influenced by output noise. On the other hand a small external disturbance can accelerate convergence rate according to the persistent excitation theory (18), small disturbances in the input $u(t)$ or in output $y(t)$ can enrich frequency of the signal $X(t)$, which is good for parameter convergence. In the following simulation we can see this point. With this prior knowledge V^0, we may start the modeling (11).

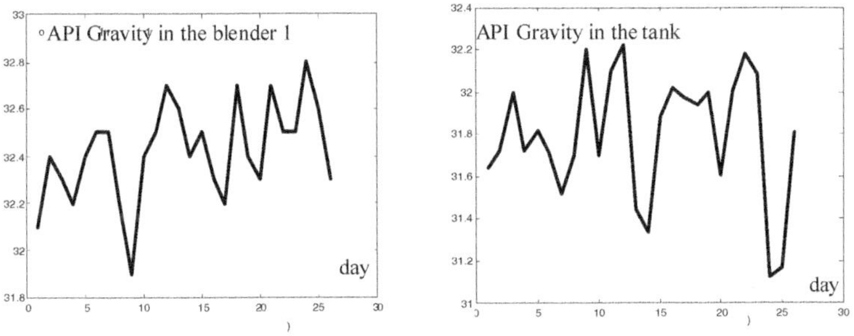

Fig. 3 The API Gravity in one month

4 Application Study

In this section, we will use real data of PEMEX and the neural networks proposed in Sect. 3 to model crude oil blending. The TMDB crude oil blending process in PEMEX is shown in Fig. 1, where the analyzers of API and flow rates are the input/output points of each block. The data are recorded in the form of Microsoft Excel daily. Each day, we have input data $[q_1, p_1, q_2, p_2, q_3, p_3, q_4, p_4]^T$ and output data $[q_f, p_f]^T$. We use "xlsread" command to transform the data sheet into Matlab. The training data are two years' records, with 730 input/output pairs. The testing data, 28 input/output pairs, are one month's records which are from another year. In this way, we can assure the testing phase is independent of the training phase. The outputs of each blender (M_1, M_2, M_3) in Fig. 1 are changed daily and different. Fig. 3 gives one month's API Gravity of two blenders.

We see that the nonlinearity of the crude oil blending is strong, and it is not easy to identify it by a simple model. We use three methods to compare the algorithm proposed in this paper, see Fig. 4.

Method A
In Fig. 4-(a) we regard the real data to satisfy the interaction model (6), *i.e.*,

$$p_f = \sum_{i=1}^{4} p_i x_i + \sum_{i=1}^{4} \sum_{j=i+1}^{4} \alpha_{i,j} x_i x_j \tag{18}$$

where $x_i = q_i/q_f$, the parameters $\alpha_{i,j}$ are linear with the data. So, we can use standard least square technique to calculate the parameter. (18) can be written as

$$p_f(k) - \sum_{i=1}^{4} p_i(k) x_i(k) = \sum_{i=1}^{4} \sum_{j=i+1}^{4} \alpha_{i,j} x_i(k) x_j(k) \tag{19}$$

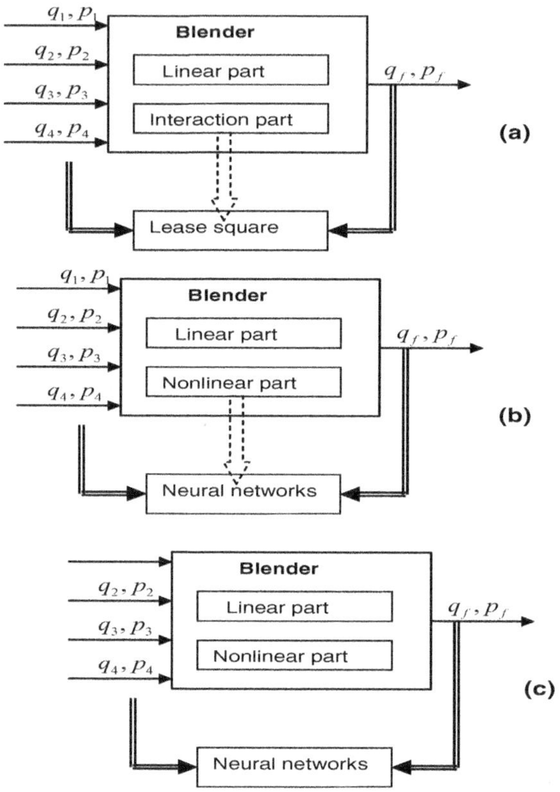

Fig. 4 Modeling crude oil blending

where $k = 1 \cdots 730$. We define

$$y(k) = p_f(k) - \sum_{i=1}^{4} p_i(k)\, x_i(k), \quad z_l(k) = x_i(k)\, x_j(k), \quad \theta_l = \alpha_{i,j}$$

where $i = 1 \cdots 4$, $j = (i+1) \cdots 4$. (19) can be written in matrix form

$$Y = Z\Theta$$

where $Y = [y(1), \ldots, y(730)]^T$, $Z = \begin{bmatrix} z_1(1) & \cdots & z_{10}(1) \\ \vdots & \ddots & \vdots \\ z_1(730) & \cdots & z_{10}(730) \end{bmatrix}^T$,
$\Theta = [\theta_1, \ldots, \theta_{10}]^T$. The least square solution is

$$\Theta = \left(Z^T Z\right)^{-1} Z^T Y$$

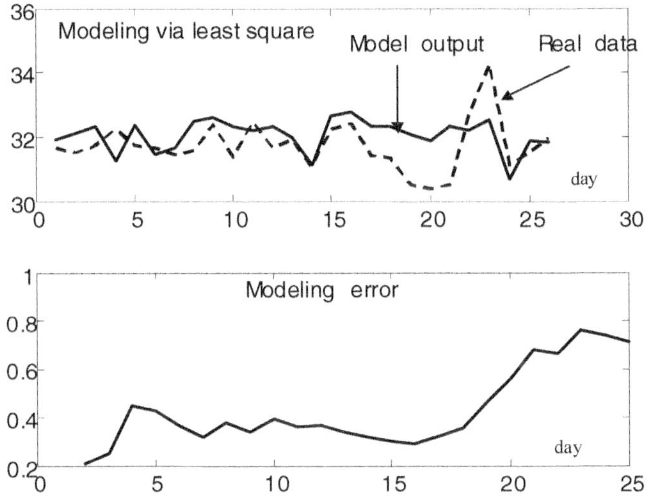

Fig. 5 Identification via interaction model and least square

After we obtain Θ (or $\alpha_{i,j}$), the other month's data are used to check the model (18). We apply the input data $[q_1, p_1, q_2, p_2, q_3, p_3, q_4, p_4]^T$ to (18). The output $\widehat{p}_f$ of the interaction model and the real data data are shown in Fig. 5.

Method B

In Fig. 4-(b) we assume the crude oil blending can be expressed as linear and nonlinear parts,

$$p_f = \sum_{i=1}^{4} p_i x_i + \Delta \tag{20}$$

It can be expressed as

$$p_f - \sum_{i=1}^{4} p_i x_i = \Delta(q_1, p_1, q_2, p_2, q_3, p_3, q_4, p_4)$$

We use following neural network model to identify Δ

$$\widehat{y}(k) = V_k \phi[W_k X(k)] \tag{21}$$

where sampling time is one day. The input to neural network is $[q_1, p_1, q_2, p_2, q_3, p_3, q_4, p_4]^T$. The output of neural network corresponds to $\left(p_f - \sum_{i=1}^{4} p_i x_i\right)$, so $X(k) = [q_1, p_1, q_2, p_2, q_3, p_3, q_4, p_4]^T$. We choose 5 nodes in the hidden layer, so $W_k \in R^{5\times 8}$, $V_k \in R^{1\times 5}$, the initial conditions for the elements of W_k and V_k are random numbers in $[0, 1]$. The modeling error is

$$e(k) = \widehat{y}(k) - \left[p_f(k) - \sum_{i=1}^{4} p_i(k)\,x_i(k)\right]$$

We use the learning algorithm (11) proposed in this paper, *i.e.*,

$$\begin{array}{l} W_{k+1} = W_k - \eta_k e(k)\,\phi' V^{0T} X^T(k) \\ V_{k+1} = V_k - \eta_k e(k)\,\phi^T \end{array} \tag{22}$$

where $\eta_k = \dfrac{s_k}{1 + \left\|\phi' V^{0T} X^T(k)\right\|^2 + \left\|\phi\right\|^2}$, $\eta = 1$, $\frac{\eta}{\pi}\overline{\zeta} = 0.2$, $\phi(\cdot) = \tanh(x) = \frac{e^x - e^{-x}}{e^x + e^{-x}}$. 730 pairs $[X(k), y(k)]$ are applied to train the neural network (21), and other 28 pairs data are used to test the training result, the modeling results are shown in Fig. 6.

Method C

In Fig. 4-(c) we consider the crude oil blending is a black-box nonlinear process, and we use neural network to model the whole system. The plant is

$$p_f = f(q_1, p_1, q_2, p_2, q_3, p_3, q_4, p_4) \tag{23}$$

The input to neural network is $[q_1, p_1, q_2, p_2, q_3, p_3, q_4, p_4]$, the output of neural network corresponds to p_f. The modeling error is

$$e(k) = \widehat{y}(k) - p_f(k)$$

We use the same neural networks and the same algorithm as in (22). We use 1500 data (four and half year´s data) to training it, after $k > 1300$, the weights are converged. Then we use another one year's flow rates of the feed stocks to test our neural model, the modeling results are shown in Fig. 7.

Using black-box identification model, three different methods in Fig. 4 give different modeling errors. We define the average modeling errors as

$$J_e = \frac{1}{n}\sum_{k=1}^{n} \left|p_f(k) - \widehat{p}_f(k)\right|$$

where $\widehat{p}_f(k)$ is the output of the models. For least square method, nonlinear part modeling and black-box modeling, J_e is 0.6, 0.025, 0.1, respectively. We have the following conclusions:

1. It is reasonable to divide the blending process into linear and nonlinear parts.
2. The interaction model (18) for the nonlinear part is not suitable in crude oil blending.
3. Neural networks and the robust learning algorithm proposed in this paper are effective for modeling of crude oil blending.

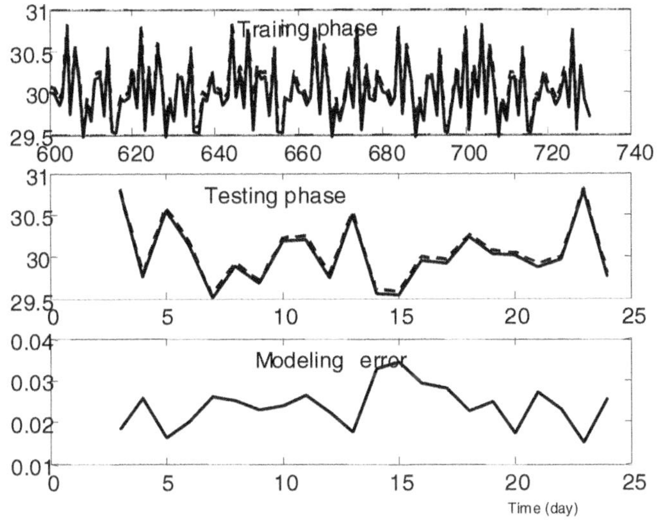

Fig. 6 Identification for nonlinear part

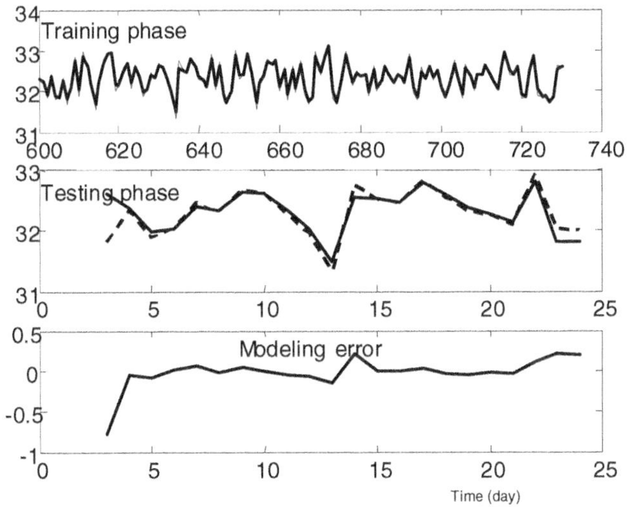

Fig. 7 Black-box identification

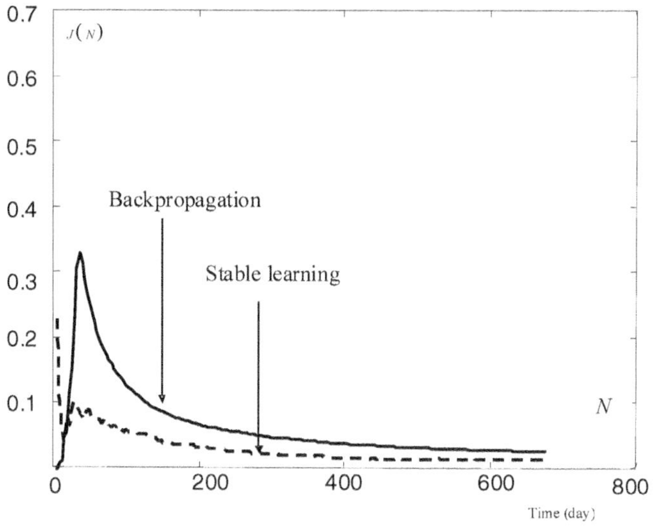

Fig. 8 Comparison of normal learning and stable learning proposed in this paper

Now we compare the dead-zone learning algorithm proposed in this paper (22) with normal backpropagation algorithm (16) in the training phase. We use the same multilayer neural networks as (16), the numbers of input layer, hidden layer and output layer are 8, 5, 1, respectively. We use a fixed learning rate $\eta = 0.05$. We found after $\eta > 0.1$ the normal backpropagation algorithm becomes unstable. The performance comparison can be realized by mean squared errors

$$J(N) = \frac{1}{2N} \sum_{k=1}^{N} e^2(k)$$

The comparison results are shown in Fig. 8. We can see that the stable algorithm proposed in this paper has a fast convergence rate, $J(730) = 0.005$. The modeling error of normal backpropagation algorithm is bigger, $J(730) = 0.078$.

5 Conclusion

In this paper a new dead-zone learning algorithm for discrete-time neural network is proposed. The theoretical analysis of stability and convergence of the neural networks are given. A real application for modeling of crude oil blending is provided. This method has great benefit for realizing model-based optimal control.

References

1. Alvarez-Ramirez, J., Morales, A., Suarez, R.: Robustness of a class of bias update controllers for blending systems. Industrial Engineering Chemistry Research 41(19), 4786–4793 (2002)
2. Brown, M., Harris, C.J.: Neurofuzzy adaptive modelling and control. Prentice-Hall, Englewood Cliffs (1994)
3. Chang, D.-M., Yu, C.-C., Chien, I.-L.: Coordinated control of blending systems. IEEE Trans. Control Systems Technology 6(4), 495–506 (1998)
4. Egardt, B.: Stability of Adaptive Controllers. LNCIS, vol. 20. Springer, Berlin (1979)
5. Feng, Z., Michel, A.N.: Robustness Analysis of a Class of Discrete-Time Systems with Applications to Neural Networks. In: Proc. of American Control Conference, San Deigo, pp. 3479–3483 (1999)
6. Gary, J.H., Handwerk, G.E.: Petroleum Refining Technology and Economics. Marcer Dekker, New York (1994)
7. Healy, W.C., Maassen, C.W., Peterson, R.T.: A new approach to blending octanes. In: Proc. 24th Meeting of American Petroleum Institute's Division of Refining, New york (1959)
8. Hopfield, J.J.: Neurons with grade response have collective computational propierties like those of a two-state neurons. Proc. of the National Academy of Science, USA 81, 3088–3092 (1984)
9. Luyben, W.L.: Process Modeling, Simulation and Control for Chemical Enqineers, 2nd edn. McGraw-Hill, Inc., New York (1990)
10. Murakami, K., Seborg, D.E.: Constrained parameter estimation with applications to blending operations. Journal of Precess Control 10, 195–202 (2000)
11. Ioannou, P.A., Sun, J.: Robust Adaptive Control. Prentice-Hall, Inc., Upper Saddle River (1996)
12. Jagannathan, S., Lewis, F.L.: Identification of Nonlinear Dynamical Systems Using Multilayered Neural Networks. Automatica 32(12), 1707–1712 (1996)
13. Jin, L., Gupta, M.M.: Stable Dynamic Backpropagation Learning in Recurrent Neural Networks. IEEE Trans. Neural Networks 10(6), 1321–1334 (1999)
14. Jiang, Z.P., Wang, Y.: Input-to-State Stability for Discrete-Time Nonlinear Systems. Automatica 37(2), 857–869 (2001)
15. Kosmatopoulos, E.B., Polycarpou, M.M., Christodoulou, M.A., Ioannou, P.A.: High-Order Neural Network Structures for Identification of Dynamical Systems. IEEE Trans. on Neural Networks 6(2), 422–431 (1995)
16. Narendra, K.S., Parthasarathy, K.: Identification and Control of Dynamical Systems Using Neural Networks. IEEE Trans. Neural Networks 1(1), 4–27 (1990)
17. Muller, A.: New method produces accurate octane blending values. Oil & Gas J. 23(3), 80–90 (1992)
18. Polycarpou, M.M., Ioannou, P.A.: Learning and Convergence Analysis of Neural-Type Structured Networks. IEEE Trans. Neural Networks 3(1), 39–50 (1992)
19. Smith, J.M., Van Ness, H.C.: Introduction to Chemical Engineering Thermodynamics. McGraw-Hill Book Company, New York (1982)

20. Singh, A., Forbes, J.F., Vermeer, P.J., Woo, S.S.: Model-based real-time optimization of automotive gasoline blending operations. Journal of Process Control 10, 43–58 (2000)
21. Song, Q.: Robust Training Algorithm of Multilayered Neural Networks for Identification of Nonlinear Dynamic Systems. IEE Proceedings - Control Theory and Applications 145(1), 41–46 (1998)
22. Suykens, J.A.K., Vandewalle, J., De Moor, B.: NLq Theory: Checking and Imposing Stability of Recurrent Neural Networks for Nonlinear Modelling. IEEE Transactions on Signal Processing (special issue on neural networks for signal processing) 45(11), 2682–2691 (1997)
23. Suykens, J.A.K., Vandewalle, J., De Moor, B.: Lur'e Systems with Multilayer Perceptron and Recurrent Neural Networks; Absolute Stability and Dissipativity. IEEE Trans. on Automatic Control 44, 770–774 (1999)
24. Yu, W., Poznyak, A.S., Li, X.: Multilayer Dynamic Neural Networks for Nonlinear System On-line Identification. International Journal of Control 74(18), 1858–1864 (2001)
25. Yu, W., Li, X.: Discrete-Time Neuro Identification without Robust Modification. IEE Proceedings - Control Theory and Applications 150(3), 311–316 (2003)
26. Yu, W.: Nonlinear system identification using discrete-time recurrent neural networks with stable learning algorithms. Information Sciences: An International Journal 158(1), 131–147 (2004)
27. Zahed, A.H., Mullah, S.A., Bashir, M.D.: Predict octane number for gasoline blends. Hydrocarbon Processing (5), 85–87 (1993)
28. Zhang, Y., Monder, D., Forbes, J.F.: Real-time optimization under parametric uncertainty a probability constrained approach. Journal of Process Control 12, 373–389 (2002)

Adaptive Self-Tuning Wavelet Neural Network Controller for a Proton Exchange Membrane Fuel Cell

M. Sedighizadeh and A. Rezazadeh

Abstract. Movable power sources of Proton Exchange Membrane Fuel Cells (PEMFC) are the key study direction in the existing Fuel Fells (FC) research field. In order to improve the performance of the PEMFC, extend its life, increase its safety and lower the system costs, it should be controlled effectively. Among various parameters to control, the electrolyte membrane moisture, affecting the performance of the PEMFC, is an important controlled variable. PEMFC presents a number of remarkable control demands. Consequently, traditional control methods cannot be applied to control PEMFC due to the imprecision, uncertainty and partial truth, and its intrinsic nonlinear characteristics. Hence, self-tuning adaptive control based on neural networks is an attractive method to control a PEMFC. In this chapter, two major wavelet neural network-based control structures will be presented. These controllers are based on the combination of a single layer feed-forward neural network with hidden nodes of adaptive wavelet functions, and an Infinite Impulse Response (IIR) recurrent network. The IIR is cascaded to the neural network to provide a new network leading to improvements in the speed of learning. These particular neural controllers assume a certain model structure to identify the system dynamics of the unknown plant approximately and generate the control signal. The first one is a self-tuning wavelet neural network controller and the second one is an adaptive self-tuning Proportional-Integral-Derivative (PID) controller using a wavelet neural network. The advantages and disadvantages of the two proposed controllers will be discussed and illustrated in this chapter. The well-known Multi Layer Perceptron (MLP) with traditional Back-Propagation training (BPP) will also be included in the control design for a base line comparison. The proposed controllers are studied in three situations: without noise, with measurement input noise, and with disturbance output noise. Finally, the results of the performance of the new controllers are compared with a multilayer perceptron network, proving a more precise modeling and control of PEMFC.

J. Schumann, Y. Liu (Eds.): Appl. of Neural Networks in High Assur. Sys., SCI 268, pp. 221–245.
springerlink.com

1 Introduction

Environmental and economical concerns on continual consumption of fossil fuels have raised demand for non-polluting and energy-efficient power generation systems. Proton Exchange Membrane Fuel Cell (PEMFC) technology has received increasing attention as the leading candidate for its non-polluting nature and high efficiency. The potential of high-energy efficiency in a PEMFC is up to 80% and only water is discharged as the waste (1).

Improving the performance of PEMFC and extending its life depend on its efficiently controlling. This system has several control parameters, but the electrolyte membrane moisture is one of the leading controlled variables. One of the challenges in attaining an optimum performance from PEMFCs is to achieve an effective hydration of the ionomeric structure of the membrane. A hydration not as much of full amount will result in a decrease in the proton conductivity of the membrane. In addition, the kinetics of the oxygen reduction reaction at the cathode is adversely affected by a decrease in water content. Due to a number of reasons, maintaining an optimum hydration level for all operating conditions of the fuel cell is difficult. For example, at low current density or at open-circuit conditions, when little liquid water is produced by the oxygen reduction reaction, the membrane hydration level will tend to drop. On the other hand, at high current densities, the electro-osmotic drag of water, associated with the proton flux from the anode to the cathode, tends to dry out the anode side of the membrane, whereas the cathode side tends to flood because of the product water generated and the water dragged across from the anode side. Hence, water management in a polymer electrolyte membrane is essential for achieving stable and high-power performance in fuel cells. Humidification of the reactant gas is one of the methods employed to keep the membrane in a high state of hydration (2).

Although many researchers have studied the mechanism of water management and have built 3D, 2D, 1D, or 0D models, few researchers have explored how to control the humidity of PEMFC stack (2). For instance, Van Nguyen et al. (4) presented a new method, which removed extra water by controlling each cell's exhaust. Chen et al. (4) introduced a humidifier system model and developed proportional feedback controller to adjust the inlet air relative humidity. Abtahi et al. (5) proposed a fuzzy controller based on Mamadani inference systems for water management of PEM. Li Guo (6) proposed a new predictive control. Yuan Ren et al. (7) introduced a predictive control based on supported vector regression machine to control humidification of hydrogen using Nafion® membrane. In (8), two methods of humidifying the anode gas, namely, external and membrane humidification are explained.

The performance of the PEMFC is dependent more on hydrogen humidification than on oxygen humidification (7); our work in this chapter is restricted to the humidification of hydrogen using a Nafion1135 membrane. Nafion1135 is a suitable and convenient humidification membrane material for fuel cell applications. One of the advantages derived from such membrane-based humidification is that, if the cooling arrangement of the stack is integrated with the humidification system, it is possible to humidify the fuel and oxidant gases at a temperature close to the operating temperature of the fuel cell (8).

When the flow rate of hydrogen, area, and type of the membrane are constant for a given PEMFC, we may change the relative humidification of hydrogen to improve the performance of the fuel cell. According to the imprecision, uncertainty and partial truth, and intrinsic nonlinear characteristics of PEMFC, traditional control methods cannot be utilized to control it. However, in this situation the self-adaptive neural network controllers are attractive methods (9). Many authors have used these controllers to control the PEMFC. For instance, Dong et al (10) developed an adaptive fuzzy control procedure for the operating temperature. Liyan Zhang et al. (11) proposed an adaptive neural control technique for PEMFC stack breathing control system. Sedighizadeh et al. (12) introduced a neural adaptive wavelet controller to control the PEMFC output power by changing of inlet reactants. In (13), Yanjun Li et al. used an artificial neural network to control the temperature fuel cells.

In this chapter, the introduced method in (12) is used to design two major wavelet neural network-based control structures. The aim of this chapter is the controlling of the output power of PEMFC by changing the humidification of hydrogen. Both of these controllers are based on a single layer feed-forward neural network with hidden nodes of adaptive wavelet function, and an Infinite Impulse Response (IIR) recurrent structure. The combination of wavelet neural network and IIR network improves the learning speed, as it provides double local structure neural network. These particular neural controllers assume a certain model structure to approximately identify the system dynamics of the unknown plant, and consequently, generate the control signal. The controllers defined in this chapter are self-tuning wavelet neural network controllers and adaptive self-tuning PID controllers using a wavelet neural network.

This chapter is organized as follows: in Section 2, a description of the employed fuel cell system model is developed. Identification algorithm and the two proposed controllers based on wavelet neural network are described in Sections 3 and 4. In Section 5, simulation results to identify the PEMFC model are presented and optimal structure for neural network is calculated. Then, simulation results to control the output power of PEMFC for the two proposed controllers are shown and the sensitivity of new controllers to input noise and output noise is estimated. At last, conclusions are made in Section 6.

2 PEMFC System Model

The key component of a PEMFC movable power source is the PEMFC stack, which is made of many PEMFC cells. The basic PEMFC cell composition includes the bipolar plate, the gas channel, the sealing gasket, the current collector, and the membrane electrode assembly (MEA). In PEMFC, MEA is the main component and thus it is called the heart of PEMFC. MEA is composed of porous gas diffusion anode, porous gas diffusion cathode, and proton exchange membrane. Its quality directly affects the PEMFC operating performance.

The PEMFC internal electrochemical reaction, in essence, is the process that combines hydrogen and oxygen over a platinum catalyst to produce water. The PEMFC mechanism is shown in Figure 1. In the PEMFC, the gases flow in gas channels of the bipolar plates and filter into the porous anode/cathode electrode layer while the humid hydrogen and oxygen enter the gas inlets.

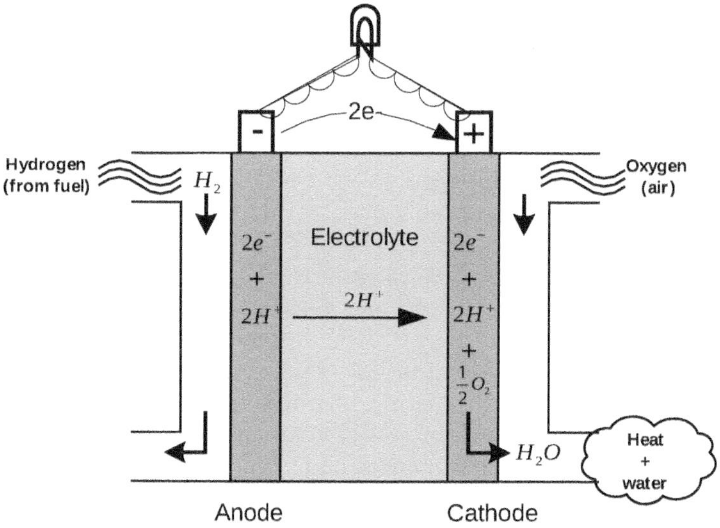

Fig. 1 Schematic diagram of PEMFC mechanism

In the anode layer, hydrogen molecules are decomposed into hydrogen ions and electrons at the presence of Platinum—the catalyst. The hydrogen ions diffuse across the proton exchange membrane and transfer to the cathode electrode layer, while the electrons flow across the load to the cathode electrode layer as the proton exchange membrane obstructs them. At the same time, in the cathode layer, oxygen molecules are decomposed into active oxygen atoms, with the aid of Platinum as the catalyst; and then react with the hydrogen ions and the electrons to produce water. Thus, PEMFC produces direct current electricity, E^0, as well as water and heat, Q^0, as byproducts (12).

PEMFC electrochemical reactions are described below:

Anode-reaction:

$$H_2 \xrightarrow{Pt} 2H^+ + 2e^- \quad (1)$$

Cathode-reaction:

$$\frac{1}{2}O_2 + 2H^+ + 2e^- \xrightarrow{Pt} H_2O + Q^o \tag{2}$$

Overall reaction:

$$\frac{1}{2}O_2 + H_2 \xrightarrow{Pt} H_2O + Q^o + E^o \tag{3}$$

As a conclusion, PEMFC can continuously generate electricity as long as hydrogen and oxygen are supplied in the definite operating condition (12).

PEMFC power sources are mainly used in portable power supplies of the electronic installation, military power supplies, and civil emergency power supplies of the transportation vehicles. Therefore, it is needed that PEMFC movable power sources be light, quickly starting, have good stability, fine performance of anti-vibration/anti-impact, and be suitable for bad conditions. Moreover, the oxidant mostly uses the air in PEMFC movable power sources, and the cooling type has the water-cooling and the air-cooling according to the generation power. In a word, PEMFC movable power sources are applied for their simplicity and practicality in order to achieve tractability, robustness, and low cost solutions. System control of PEMFC movable power source belongs to the complex electrochemical process control (12). Firstly, the system control characteristics of PEMFC movable power source are multi- input-output and nonlinear. In real time, many variables need to be controlled in the PEMFC movable power source, such as moisture, temperature, pressure, flux, load, and so on. Furthermore, the mapping of the input and output variables has an extremely complex nonlinear relation and changing of disturbance factors intensifies this nonlinear status. Secondly, PEMFC movable power system control has the characteristics of time changing, distribution parameter, and strong coupling. Subsequently, system control of PEMFC movable power source has the characteristics of long-hysteresis and restraint. The controlled object has a big lag for the reason of the electrochemical reaction. Eventually, system control of PEMFC movable power source has the complex characteristics of uncertainty and random disturbance. The control process has the dynamic uncertainty owing to the changes of the reactant and the internal status of PEMFC stack. In addition, there are bad qualities of the random disturbance and the transmission noise in PEMFC stack. Various attempts are being made to model fuel cell systems. According to the PEMFC dynamic characteristics, the output power model of the PEMFC stack can be described as (6):

$$\dot{P}(t) = F[P(t), T(t), V_w(t), V_a(t), V_c(t), M_h(t), M_o(t)] \tag{4}$$

We denote the operating temperature of the stack by $T(t)$, the speed of the cooling water by $V_w(t)$, the speed of the hydrogen by $V_a(t)$, the speed of the air by $V_c(t)$, the humidification of hydrogen by $M_h(t)$, the humidification of oxygen by $M_o(t)$, and the output power of the stack by $P(t)$. The rate of change of power of the stack has a nonlinear relation, $F(\cdot)$ of the foregoing variables as in (4).

To facilitate the modeling and control design, we use the following discrete form of Eq. (4):

$$P(k+1) = F[P(k), T(k), V_w(k), V_a(k), V_c(k), M_h(k), M_o(k)] \quad (5)$$

Since the performance of the PEMFC is dependent more on hydrogen humidification than on oxygen humidification (8), our work in this chapter is restricted to the humidification of hydrogen using Nafion1135 membrane. While type of the membrane, operating temperature, area, and the flow rate of hydrogen, oxygen, and cooling water are constant for a given PEMFC, we may change the relative humidification of hydrogen to improve performance of the fuel cell (8). The relationship between power output of the fuel cell and humidification of hydrogen can be simplified as a first-order unknown difference equation with time delay (6):

$$P(k+1) = F[P(k), M_h(k)] \quad (6)$$

or

$$y(k+1) = F[y(k), u(k)] \quad (7)$$

where $y(k)$ is the output power of PEMFC, and u is the system input, which is the relative humidification of hydrogen, $M_h(k)$. Based on these characteristics, the traditional control method of PEMFC movable power source is not often able to meet the effective control result.

3 Wavelet Neural Network and Identification Algorithm

3.1 Wavelet Neural Network

In order to deal with the tracking operation using a neural network based controller, the unknown nonlinear PEMFC should be identified according to a particular model. In this particular identification process, the model consists of a neural network topology similar to a Radial Basis function network topology with the wavelet transform used in the hidden units. This network is called "wavenet". The concept of wavenet introduces a super-wavelet, i.e., a wavelet that is a linear combination of daughter wavelets. Daughter wavelets are simply a dilated and shifted version of the original mother wavelet. The super-wavelet allows the shape of the wavelet to adapt to a particular problem—a concept that goes beyond adapting the parameters of a fixed shape wavelet. This network has shown good results in nonlinear systems identification and control (14). A local infinite impulse response (IIR) block structure is cascaded with the network. The IIR synopsis network is used to create the double local network architecture. This architecture provides a computationally efficient method for training the system, resulting in quick learning and fast convergence (14). The algorithm of proposed neural network adaptive wavelets is similar to that in (14), where any desired signal $y(t)$ can be modeled by a linear combination of Morlet daughter wavelets, $h_{a,b}(t)$. Here, $h_{a,b}(t)$ is generated from a Morlet mother wavelet by dilation, a, and translation, b, as follows:

$$h_{a,b}(t) = h\left(\frac{t-b}{a}\right) = \cos\left[\omega_o\left(\frac{t-b}{a}\right)\right]\exp\left[-0.5\cdot\left(\frac{t-b}{a}\right)^2\right] \quad (8)$$

with the dilation factor $a > 0$. ω_o is the wavelet frequency, which chosen as $\omega_o = 4$, as it meets approximately the admissibility condition (14).

The approximated signal of the network $\hat{y}(t)$ can be modeled as (14):

$$\hat{y}(t) = \sum_{i=0}^{M} c_i z(t-i)u(t) + \sum_{j=1}^{N} d_j \hat{y}(t-j)v(t) \quad (9)$$

where,

$$z(t) = \sum_{k=1}^{K} w_k h_{a_k,b_k}(t) \quad (10)$$

In (9) and (10), k is the number of wavelets, and w_k is the kth weight coefficient; M is the number of feed-forward delays, and c_j is the feed-forward coefficient of the IIR filter. N is the number of feedback delays, and d_j is the recursive filter coefficients. The signals $u(t)$ and $v(t)$ are the input (the relative humidification of hydrogen), and co-input to the system at time t, respectively. Input $v(t)$ is usually kept small for feedback stability purposes (14).

The neural network parameters a_k, b_k, c_i, w_k, and d_j can be calculated by minimization of the least mean square (LMS) of output error. The objective function is:

$$E = \frac{1}{2}\sum_{t=1}^{T} e^2(t) \quad (11)$$

Where, error $e(t)$ is the time varying error function. $y(t)$ is the desired response or the PEMFC output power and $\hat{y}(t)$ is the output of the neural network.

$$e(t) = y(t) - \hat{y}(t) \quad (12)$$

To minimize the cost function, we may use the method of steepest descent which requires the gradients $\frac{\partial E}{\partial w_k}$, $\frac{\partial E}{\partial b_k}$, $\frac{\partial E}{\partial a_k}$, $\frac{\partial E}{\partial c_i}$, and $\frac{\partial E}{\partial d_j}$ for updating the incremental changes of each particular parameter w_k, b_k, a_k, c_i, and d_j, respectively. For Morlet mother wavelet, gradients of E are:

$$\frac{\partial E}{\partial w_k} = -\sum_{t=1}^{T} u(t)e(t)\sum_{i=0}^{M} c_i h(\tau - i) \tag{13}$$

$$\frac{\partial E}{\partial b_k} = -\sum_{t=1}^{T} u(t)e(t)\sum_{i=0}^{M} c_i w_k \frac{\partial h(\tau - i)}{\partial b_k} \tag{14}$$

$$\frac{\partial E}{\partial a_k} = -\sum_{t=1}^{T} u(t)e(t)\sum_{i=0}^{M} c_i \tau w_k \frac{\partial h(\tau - i)}{\partial b_k} = \tau \frac{\partial E}{\partial b_k} \tag{15}$$

$$\frac{\partial E}{\partial c_i} = -\sum_{t=1}^{T} u(t)e(t)z(t - i) \tag{16}$$

$$\frac{\partial E}{\partial d_j} = -\sum_{t=1}^{T} v(t)e(t)\hat{y}(t - i) \tag{17}$$

where, $\tau = \frac{t-b_k}{a_k}$ and we have:

$$\frac{\partial h_{a,b}(t)}{\partial b} = \frac{1}{a}\left[\omega_o \sin\left[\omega_o\left(\frac{t-b}{a}\right)\right] \exp\left[-0.5 \cdot \left(\frac{t-b}{a}\right)^2\right] + \left(\frac{t-b}{a}\right) h_{a,b}\left(\frac{t-b}{a}\right)\right]$$

The incremental changes of each parameter are simply the negative of their gradients,

$$\Delta w = -\frac{\partial E}{\partial w}, \Delta b = -\frac{\partial E}{\partial b}, \Delta a = -\frac{\partial E}{\partial a}, \Delta c = -\frac{\partial E}{\partial c}, \Delta d = -\frac{\partial E}{\partial d} \tag{18}$$

Thus, each coefficient vector of the network, i.e., $\mathbf{w}$, $\mathbf{b}$, $\mathbf{a}$, $\mathbf{c}$, and $\mathbf{d}$ is updated in accordance with the following rules:

$$\mathbf{w}(n+1) = \mathbf{w}(n) + \mu_w \Delta\mathbf{w} \tag{19}$$

$$\mathbf{b}(n+1) = \mathbf{b}(n) + \mu_b \Delta\mathbf{b} \tag{20}$$

$$\mathbf{a}(n+1) = \mathbf{a}(n) + \mu_a \Delta\mathbf{a} \tag{21}$$

$$\mathbf{c}(n+1) = \mathbf{c}(n) + \mu_c \Delta\mathbf{c} \tag{22}$$

$$\mathbf{d}(n+1) = \mathbf{d}(n) + \mu_d \Delta\mathbf{d} \tag{23}$$

where, the values of μ_x coefficients are fixed learning rate parameters.

3.2 System Model Identification

Consider a general single-input single-output (SISO) dynamic system, similar to (6), represented by the following state equations:

$$\dot{x} = f(x(t), u(t), t) \tag{24}$$

$$y(t) = g(x(t), t) \tag{25}$$

The equations (24) and (25) can be written in discrete time space as:

$$\begin{aligned} x(k+1) &= f(x(k), u(k), k) \\ y(k) &= g(x(k), k) \end{aligned} \tag{26}$$

where, $x(k) \in R^n$, and $u(k), y(k) \in R$. The only available data are the input u and the output y. If the linear behavior of the system around the equilibrium state is noticeable, then an input-output representation exists, which has the following form:

$$y(k+1) = \varphi(y(k), y(k-1), \ldots, y(k-n+1), u(k), u(k-1), \ldots, u(k-n+1)) \tag{27}$$

The $\varphi(\cdot)$ function maps $y(k)$, $u(k)$, and their $n-1$ past values onto $y(k+1)$. In this light, a neural network model $\hat{\varphi}$ can be trained to approximate φ over the interest domain. Practically if an exact model of the plant were available, approximate models would be adapted to update the control parameters. The alternative model of an unknown plant, that can simplify the computation of the control input, is described by the following equation:

$$\begin{aligned} y(k+1) = {} & \\ & \phi(y(k), y(k-1), \ldots, y(k-n+1), u(k), u(k-1), \ldots, u(k-n+1)) + \\ & \Gamma(y(k), y(k-1), \ldots, y(k-n+1), u(k), u(k-1), \ldots, u(k-n+1))u(k) \end{aligned} \tag{28}$$

As the system in (6) is first order, we can express the above equation as follows:

$$y(k+1) = \phi(y(k)) + \Gamma(y(k))u(k) \tag{29}$$

where, $y(k)$ and $u(k)$ are the output and the input at the kth instance of time respectively.

If the nonlinearity terms $\phi(\cdot)$ and $\Gamma(\cdot)$ are exactly known, the required control $u(k)$ for tracking a desired output $r(k+1)$ can be computed at every time instance using the following formula:

$$u(k) = \frac{r(k+1) - \phi(y(k))}{\Gamma(y(k))} \tag{30}$$

However, if $\phi(\cdot)$ and $\Gamma(\cdot)$ are unknown, the idea is to use the neural network adaptive wavelets model to approximate the system dynamics, i.e.:

$$\hat{y}(k+1) = \hat{\phi}(y(k), \Theta_\phi) + \hat{\Gamma}(y(k), \Theta_\Gamma)u(k) \tag{31}$$

Comparing the model of (31) with (9), we can conclude:

$$\hat{\phi}(y(k), \Theta_\phi) = \sum_{j=1}^{N} d_j \hat{y}(k-j)v(k) \tag{32}$$

$$\hat{\Gamma}(y(k), \Theta_\Gamma) = \sum_{i=0}^{M} c_i z(k-i) \quad (33)$$

4 Proposed Controller Design

4.1 Neural Network Controller Based on Wavelet

The nonlinearities of $\phi(\cdot)$ and $\Gamma(\cdot)$ are approximated by two distinct neural network functions, $\hat{\phi}(\cdot)$ and $\hat{\Gamma}(\cdot)$, with adjustable parameters (including weights w_k, dilations a_k, translations b_k, IIR feed-forward coefficients c_k, IIR feedback coefficients d_k) represented by Θ_ϕ and Θ_Γ, respectively. The control $u(k)$ for tracking a desired output $r(k+1)$ can be obtained from the following equation:

$$u(k) = \frac{r(k+1) - \hat{\phi}(y(k), \Theta_\phi)}{\hat{\Gamma}(y(k), \Theta_\Gamma)} \quad (34)$$

The neuro controller for self-tuning control PEMFC is provided in Figure 2.

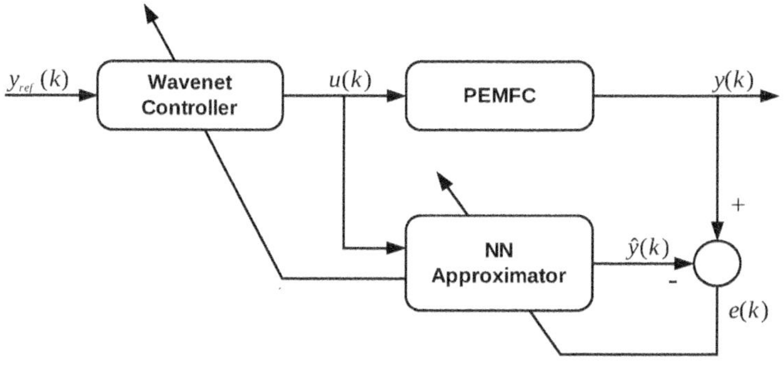

Fig. 2 Block diagram of closed loop wavenet control

4.2 PID Neural Network Controller Based on Wavelets

The PID control $u(k)$ for tracking a desired output $r(k+1)$ can be obtained from (15):

$$u(k) = u(k-1) + P[\epsilon(k) - \epsilon(k-1)] + I\epsilon(k) + D[\epsilon(k) - 2\epsilon(k-1) + \epsilon(k-2)] \quad (35)$$

where, P, I, and D are proportional, integral, and differential gains, respectively. $u(k)$ is the plant input at time KT and T is the sampling interval, and:

$$\epsilon(k) = r(k) - y(k) \quad (36)$$

P, I, and D parameters are considered as a part of the function E and can updated according to the optimization of the cost function, E, of Eq. (11), as following:

$$P(k) = P(k-1) + \mu_P e(k)\Gamma(k)(\epsilon(k) - \epsilon(k-1)) \quad (37)$$
$$I(k) = I(k-1) + \mu_I e(k)\Gamma(k)\epsilon(k) \quad (38)$$
$$D(k) = D(k-1) + \mu_D e(k)\Gamma(k)(\epsilon(k) - 2\epsilon(k-1) + \epsilon(k-2)) \quad (39)$$

where, $\hat{\Gamma}$ is derived from Eq. (33), and μ_x is the fixed learning rate of each adaptive PID parameter. Founded on the PID controller for self-tuning PEMFC's controls, Figure 3 depicts the block diagram of the resulting network topology.

The optimum output power P_{opt} is used as a reference value for the closed loop control of PEMFC. The characteristics of the PEMFC used for simulations and experiments in this chapter, are shown in Table 1, but they are considered as unknown for the controller. For this reason, the number of wavelets was obtained on a trial-and-error basis.

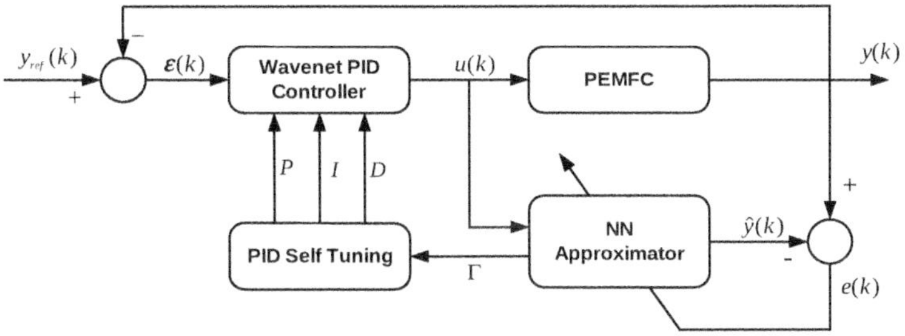

Fig. 3 Block diagram of closed loop PID control

Table 1 The parameters of control system of the 1kW PEMFC power test

Item	Value
Number of single cells	42
Available area of single cell	235 cm^2
Pt content of cathode/anode	0.5/0.5 gcm^{-2}
Proton exchange membrane	Nafion 1135
Operating temperature of the stack	$50°C - 100°C$
Pressure of the flowing air of the stack	0.28 MPa
Pressure of the flowing hydrogen of the stack	0.1 MPa
Pressure of the cooling water	0.09 MPa
Flow rate of the fuel cell	12–20 L/min
Flow rate of the cooling water	0–100 L/h

5 Simulation Results

5.1 *Identification of PEMFC*

Performing various tests on a 1KW PEMFC movable power source, the input-output data required for training of neural network, is derived. The data is applied to wavelet-network based neural adaptive controllers in order to simulate and obtain the best control effects in typical working conditions. Using the PEMFC data extracted from the tests, the *wavenet* network with different sizes of Morlet mother wavelets is used to identify the PEMFC model. Furthermore, the IIR block structure with four feed-forward delay blocks and four feedback blocks is employed. Wavelets are local basic functions, providing less interference than global ones. This results in a noncomplex dependency in the parameters of neural network (14).

The aforementioned idea will now be confirmed by presenting a number of clarifications derived from the results of the MATLAB simulations. With the assumption that the training data are stationary and rich enough, optimal performance with a small learning rate is usually achievable. Therefore, all of the learning rate parameters for weights, dilations, translations, IIR feed-forward coefficients, and feedback coefficients are set to 0.006, 0.027, 0.027, 0.012, and 0.012, respectively; and all initial weights w_k and dilations a_k are fixed at zero and five, correspondingly. As soon as the desired normalized error of 0.03 is reached, learning will terminate. The following simulations demonstrate the results of the *wavenet* network performance when utilizing the Morlet super-mother wavelet. To conclude, we can say that a *wavenet* network composed of more wavelets can quickly get to initial convergence with reference to the number of iterations. However, to reach the desired error goal of 0.03, networks having many wavelets cannot converge easily and the error begins to oscillate. Choosing the step size of learning rate in a large scale will cause the iteration process to jump between two opposite sides of a valley instead of following the natural gradient contour. Figure 4 shows that, when the number of wavelets K is small, e.g., $K = 4$, reaching the error of 0.85 takes 45 iterations while for $K = 20$ it takes 90 iterations to reach the error of 0.35. On the other hand, when the error of 0.03 is the target, $K = 6$ takes 37 iterations while $K = 25$ takes 705 iterations. In addition, large K are undesirable because in this case there are more coefficients to be updated. Small K can also take a great deal of time to be calculated; e.g., for $K = 4$, the error of 0.03 is reached after more than 1203 iterations. The number of iterations to reach the desired error for various numbers of wavelets is illustrated in Table 2 and Figure 4. In conclusion, setting the number of Morlet wavelets as K = 8 is satisfactory to identify the unknown PEMFC model, because the minimum number of iteration for specified error target is met by this number of wavelets. To model the Wavenet identifiers, it was obtained that more hidden units can not enhance performance, nonetheless may decline the network's stability. The minimum number of wavelets

Table 2 Number of iterations computed in error vs. number of employed Morlet wavelets

Error	Number of Wavelets					
	4	6	8	12	20	25
0.85	45	13	**10**	14	17	22
0.35	120	25	**13**	37	90	120
0.05	570	32	**17**	42	175	250
0.03	1203	37	**24**	57	345	705

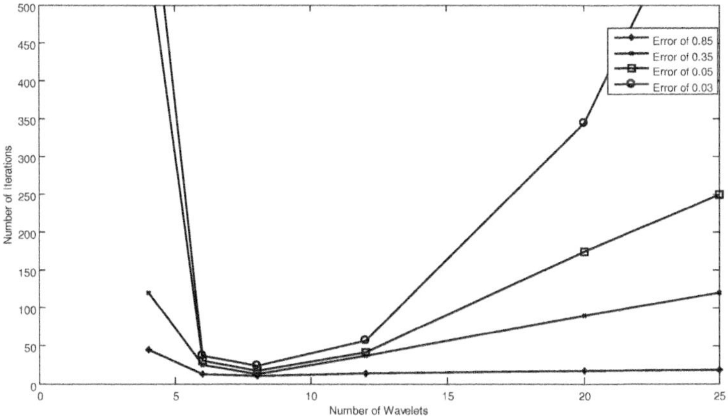

Fig. 4 Iterations vs. Number of Morlet Wavelets per Normalized Errors

(hidden units), depends on the types of wavelets and sizes of data. To specify the minimum number of wavelets in network can be an open research subject. However it is clear that the network with less number of parameters is more efficient network. In this chapter, the minimum number of wavelets was approximately 8 wavelets.

5.2 Control of PEMFC without Noise

5.2.1 Neuro *Wavenet* Controller

According to Figure 2, later than identification of PEMFC, an output power reference signal is applied to the system. Figure 5 shows the results of the tracking of output power reference using the proposed self-tuning neural wavenet controller with eight Morlet wavelets. In the developed controller, the IIR block structure utilizes four feed-forward delay blocks and four feedback blocks. Curves (2) and (3) in Figure 5 represent output power response from actual PEMFC plant and neural network identifier, respectively, when the same control is fed to both of them. The mean squared error (MSE) with respect to output reference (Curve (1)) of actual plant and neural network responses are 0.017 and 0.0067, respectively. It is obvious that unknown PEMFC and identifier responses follow the reference fairly well; but the neural-network output track reference signal is faster than actual plant response. Figure 6 displays the updating of parameters. It should be noticed that during the tracking period super wavelet parameters (w_k, a_k, and b_k) have smaller variations than the IIR coefficients c_k, and d_k. This reveals that after the identification process of the *wavenet* network, the network parameters become very stable around the operation point. Figure 6b shows the updating of the nonlinearity terms $\hat{\Phi}(k)$ and $\hat{\Gamma}(k)$ during the tracking period.

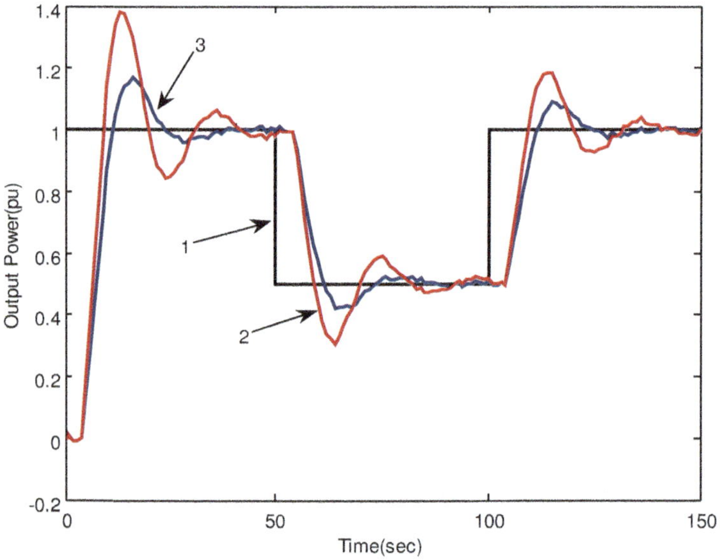

Fig. 5 Self-Tuning Neuro *Wavenet* Controller Responses to PEMFC output power reference. 1-Setpoint Reference 2- Plant Output Response 3- NN Output Response

5.2.2 Adaptive PID Controller Using *Wavenets*

In accordance with the controller block diagram in Figure 3, Figure 7 demonstrates the tracking of reference signal by actual plant and neural network identifier, respectively, using the proposed adaptive self-tuning PID controller with self-tuning *wavenet* parameter adaptation. The neural network structure and IIR structure are equal with the previous controller. Super mother wavelets parameters and IIR coefficients are initiated similar to the previous section. The proportional (P), integral (I), and derivative (D) gains of the controller are initially set at 3, 0.2, and 0.01, respectively, which then

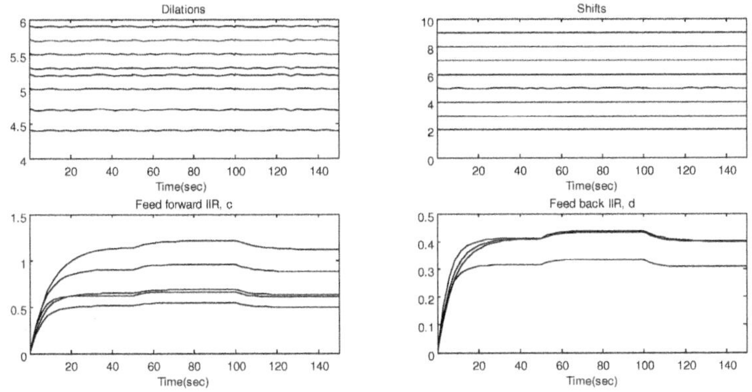

(a) Neuro *Wavenet* Parameters

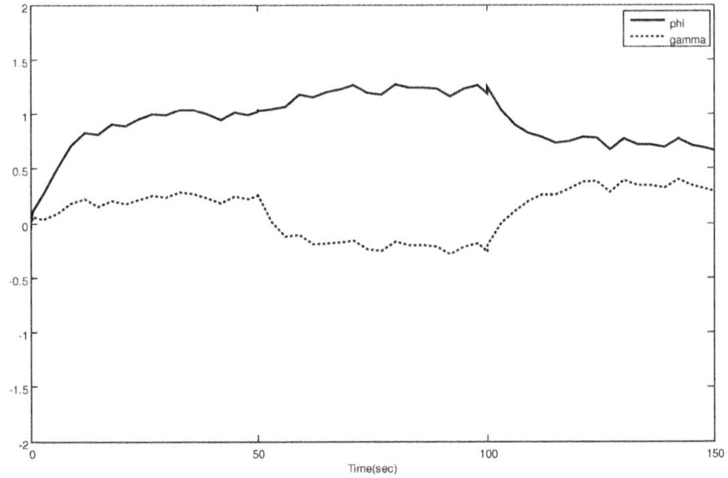

(b) Nonlinearity Term Updates $\hat{\Phi}(k)$ and $\hat{\Gamma}(k)$

Fig. 6 Neuro *Wavenet* Parameters

will vary with control neural network conditions. Once more, the emphasis is on the plant output responses to the reference set point. In contrast with the previous controller, the neural network response follows output reference faster than the PEMFC response, in the proposed adaptive PID control; the actual plant output tracks the desired output faster than the neural network responses. The MSE values, concerning the desired output power of the actual plant and of the neural network output response, are 0.036 and 0.07, respectively. Figure 8 illustrates adaptive updates to all parameters applied to the control scheme.

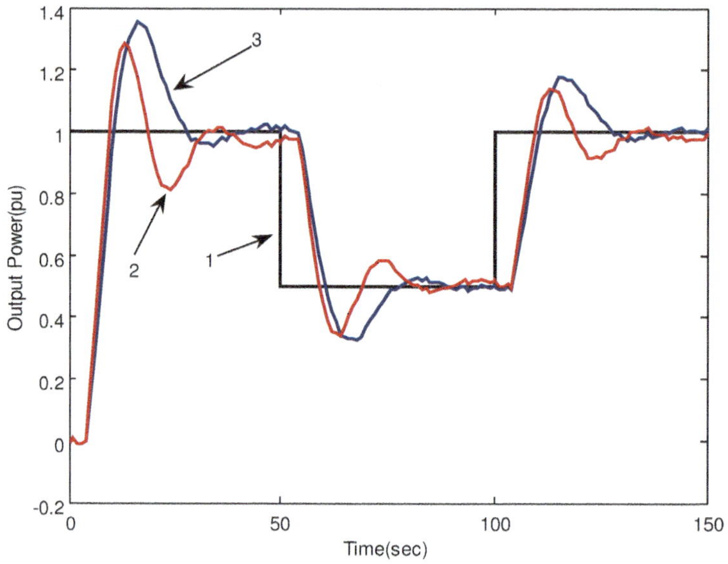

Fig. 7 Adaptive Self-Tuning PID Controller Responses to Set-Point Reference. 1-Setpoint Reference 2-Plant Output Response 3-NN Output Response

5.2.3 Neuro-Controller Based on the Back-Propagation Algorithms

In this section, a traditional feed-forward MLP neural network structure, based on back-propagation algorithms as a baseline for comparison, is introduced. This two-layer perceptron network, with one hidden layer and one output layer, has eight hidden nodes. For simulation, the bias for each node is set at 0.4 and initial weights of both layers are set at random. Adaptive learning rate and its momentum coefficient are initially set at 0.005 and 0.97, respectively, and activation functions in hidden units are hyperbolic tangents. We know that the back-propagation learning does not fulfill a well on-line control because of its global characteristics. Figure 9 illustrates the results of tracking of the reference output using the Back Propagation (BPP) algorithm. Considering the PEMFC output taken to be 0.14, it needed a longer time to reach the target output and mean square error. In this figure, it is shown that the neural network output cannot pick up the desired output and is not acceptable for this particular control application.

5.3 Control of PEMFC with Input Noise

The block diagram of PEMFC control with input noise is similar to the block diagrams of Figures 2 and 3 with measurement noise, added to the input or

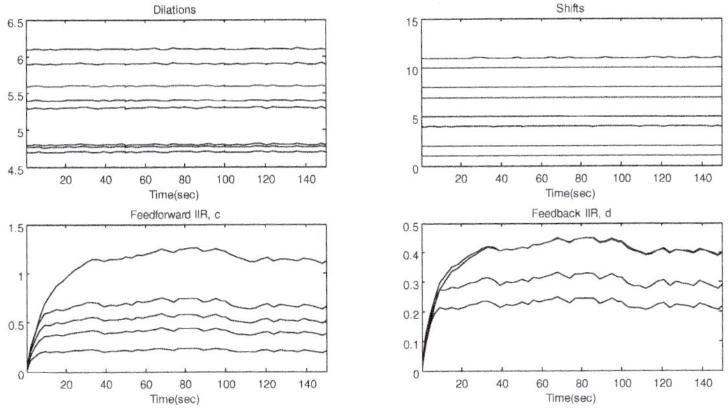

(a)

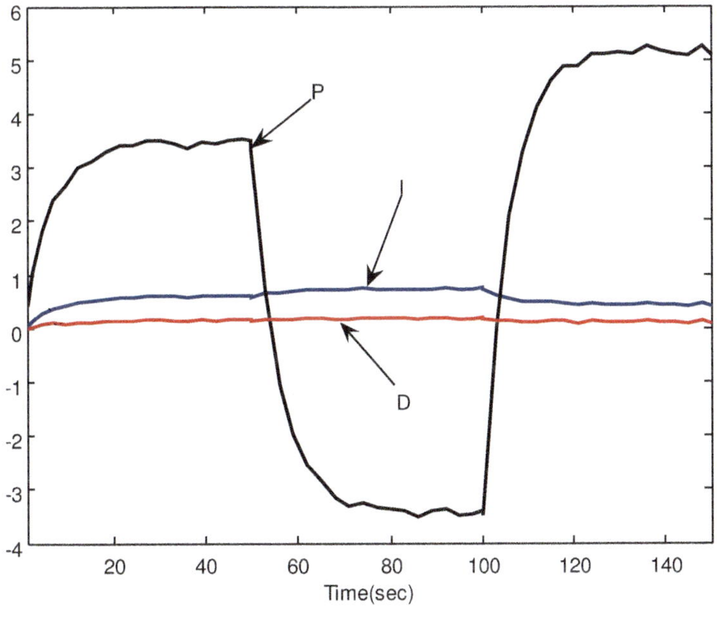

(b)

Fig. 8 Adaptive Self-Tuning PID Parameter Updates

output of the neural network block. There is no difference in applying the noise to the input or output of NN. This equivalence can easily be represented

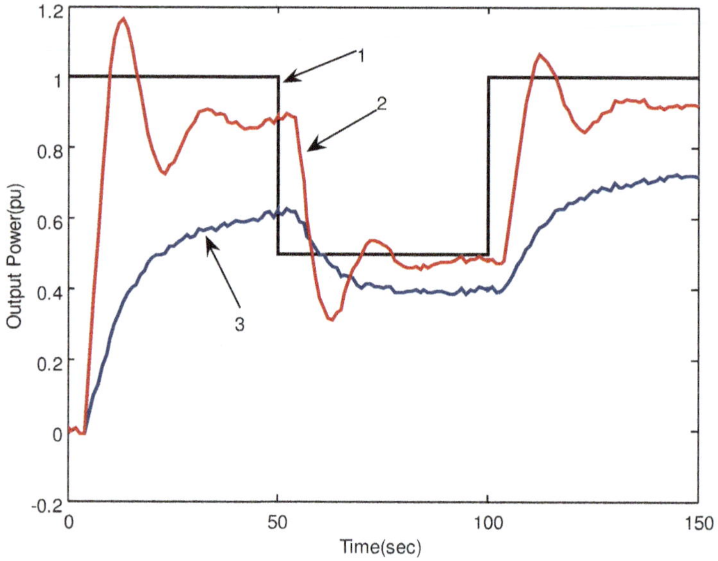

Fig. 9 Neuro-Controller with BPP Learning Responses. 1-Setpoint Reference 2- Plant Output Response 3- NN Output Response

by Equations (29) and (31). When the noisy input $w(k)$ is added to the input port of network, the wavenet output response is given by:

$$\begin{aligned}\hat{y}(k+1) &= \hat{\Phi}(k) + \hat{\Gamma}(k)(u(k) + w(k)) \\ &= \hat{\Phi}(k) + \hat{\Gamma}(k)u(k) + \hat{\Gamma}(k)w(k) \\ &= \hat{\Phi}(k) + \hat{\Gamma}(k)u(k) + \hat{w}(k)\end{aligned} \tag{40}$$

The *wavenet* response to the noise, added to the network output, is represented by:

$$\hat{y}(k+1) = \hat{\Phi}(k) + \hat{\Gamma}(k)u(k) + w(k) \tag{41}$$

It is observed that (40) and (41) provide the same results to the identification performance. Once the identification process for the unknown nonlinear PEMFC system has been performed, control action is activated to follow the desired reference. The measurement noise $w(k)$ of random distribution with a variance of 0.01 is applied at the input port of the *wavenet*. Simulations of each control methods are shown in the next subsections.

5.3.1 Neuro *Wavenet* Controller

All parameters stay the same as in the Section 5.2.1, Figure 10 represents the tracking of reference signal by actual plant and neural network identifier, respectively. According to the results, it is clear that although the measurement

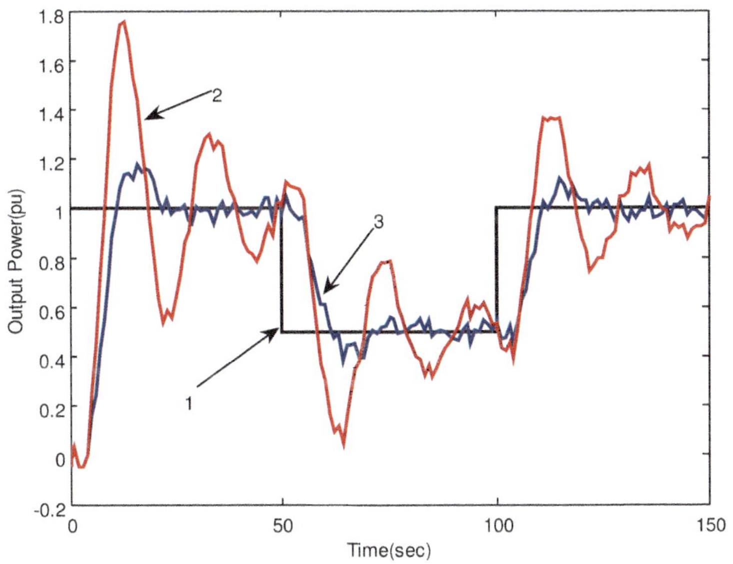

Fig. 10 *Wavenet* Controller Responses to Set-Point Control with input noise. 1-Setpoint Reference 2- Plant Output Response 3- NN Output Response

noise is only inserted to the input of neural network, the plant response to the control action is noisy. From this figure, it can be shown that the neural network response is better than that of the actual plant in tracking the desired reference to the corresponding MSE of 0.03 and 0.07, respectively. It can be concluded that in this control scheme, the input noise creates the main problem for tracking of the set point for the actual system.

5.3.2 Adaptive PID Controller Using *Wavenets*

Figure 11 shows the simulation results of the set-point control using the proposed adaptive self-tuning PID controller. The neural network structure is similar to Section 5.2.2 and all of parameters are initiated with same values in Section 5.2.2. In the proposed control method, the plant output responds to the desired set point much better and quicker than in the previous control algorithms. The MSE for the actual plant and the NN output with respect to the reference are 0.02 and 0.35, respectively. The control action to the actual plant is now less noisy and with less overshoot than the previous method. The NN response to the noisy control varied slowly and did not reach the set point.

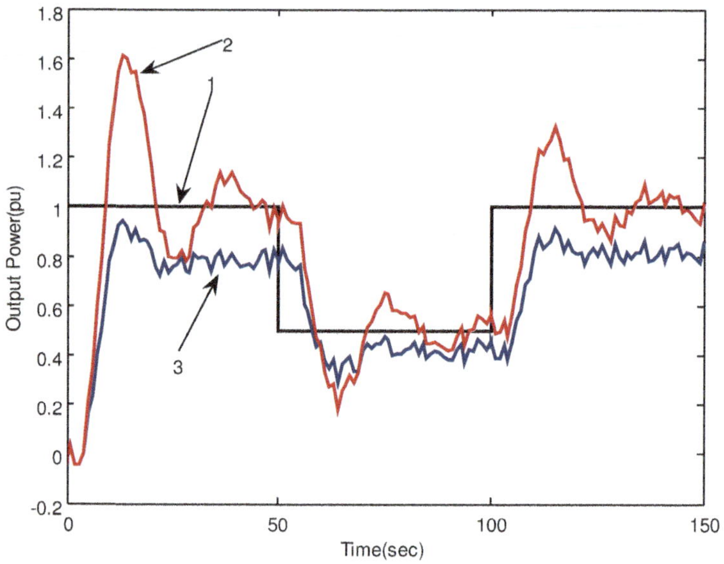

Fig. 11 PID Controller Responses to Set-Point Control with input noise. 1-Setpoint Reference 2-Plant Output Response 3-NN Output Response

5.3.3 Neuro-Controller Based on the Back-Propagation Algorithms

With the same initialization as in Section 5.2.3, the simulation result of the input noise study for the two-layer feed-forward neural network with back-propagation algorithms is shown in Figure 12. The MSE of the plant was found to be 0.06. The neural network output hardly responds to the tracking operation, and the plant output is noisy. Unlike *wavenets*, the noise seriously affects the responses of Back Propagation (BPP) algorithm.

5.4 Control of PEMFC with Output Noise Problem

The block diagram of an output noise model can be equivalently interpreted as the same block diagrams of Figure 2 and 3 with the disturbance output noise added to the input-output port of the plant. In (42) and (43), it is shown that we can use the input noise to the plant rather than modeling the noise in the plant output. When the model, in which the noise added to output of the plant is used, the error becomes:

$$\begin{aligned} e(k) &= (y(k) + w(k)) - \hat{y}(k) \\ &= [\Phi(k) - \hat{\Phi}(k)] + [\Gamma(k)u(k) - \hat{\Gamma}(k)u(k)] + w(k) \end{aligned} \tag{42}$$

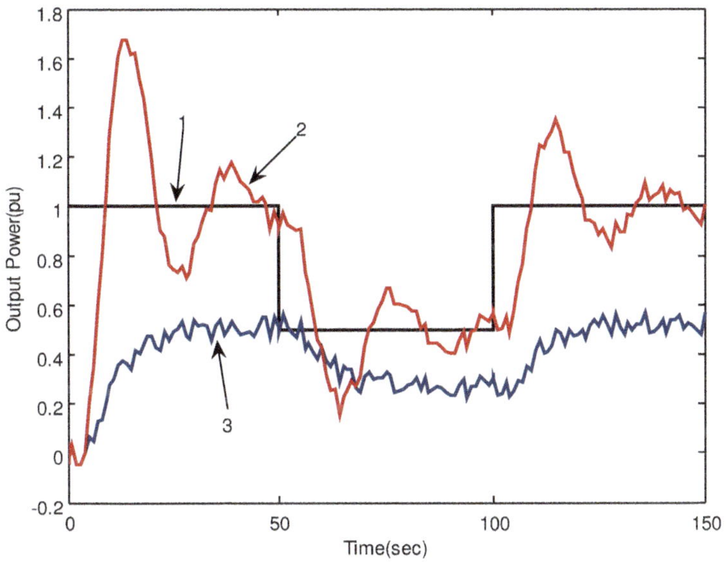

Fig. 12 Neuro-Controller with BPP Learning Responses with input noise. 1-Setpoint Reference 2-Plant Output Response 3-NN Output Response

In addition, the error of the model with input noise added to the plant is:

$$\begin{aligned} e(k) &= [\Phi(k) - \hat{\Phi}(k)] + [\Gamma(k)u(k) - \hat{\Gamma}(k)u(k)] + \Gamma(k)w(k) \\ &= [\Phi(k) - \hat{\Phi}(k)] + [\Gamma(k)u(k) - \hat{\Gamma}(k)u(k)] + \hat{w}(k) \end{aligned} \quad (43)$$

Thus, the two identification models provide similar algorithmic approximations. The disturbance noise $w(k)$, with a variance of 0.01 is inserted at the input port of the nonlinear plant PEMFC.

5.4.1 Neuro *Wavenet* Controller

All parameters stay the same as in the Section 5.2.1. Figure 13 represents output plant and neural network responses to the tracking of set point, respectively. According to this figure, there is an offset in the plant output response to the reference value, due to the disturbance at the output port of the plant. Initial oscillation to both plant and neural network responses is guessed to originate from the stability of the neural network adaptation caused by the scaling to the input port of the neural network. With this control scheme, the neural network performance is noisy, but quick to the set point, while the plant output performance is affected by the disturbance of the noise.

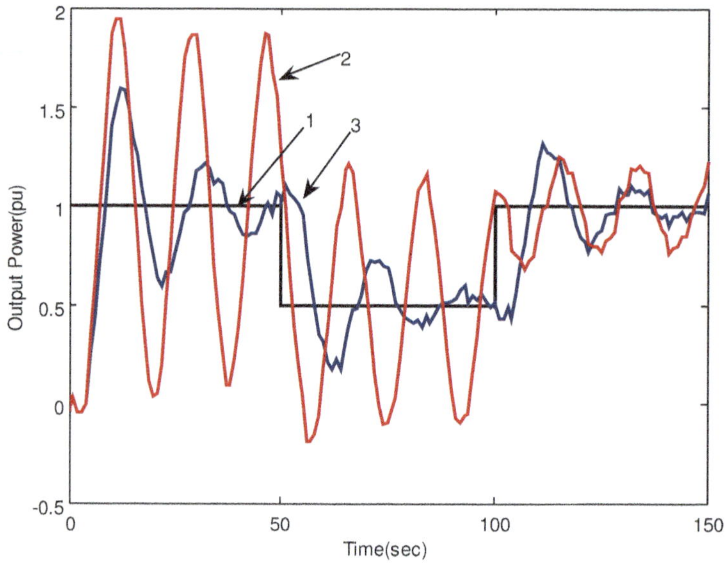

Fig. 13 Wavenet Controller Responses to Set-Point Control with output noise. 1-Set point 2-Plant Output Response 3-NN Output Response

5.4.2 Adaptive PID Controller Using *Wavenets*

Figure 14 shows the simulation results to the set-point control using the proposed adaptive self-tuning PID controller. The neural network structure is similar to Section 5.2.2 and all of the parameters are initiated with same values in Section 5.2.2. Again, the plant output response is noisy but achieves the set point tracking while the NN response is only used for the P, I and D adaptation processes. The MSE of the actual plant and of the *wavenet* network, found with respect to the desired reference, were 0.04 and 0.45, respectively.

5.4.3 Neuro-Controller Based on the Back Propagation Algorithms

With the same initialization as in Section 5.2.3, the simulation result of the output noise study for the two-layer feed-forward neural network with back-propagation algorithms is shown in Figure 15. The plant responds with the MSE of 0.06. In this conventional algorithm, the plant response is noisy with a small offset and the neural network responses never adjust to the changes. Finally, the output noise studies show the worst scenario in the adaptive self-tuning control.

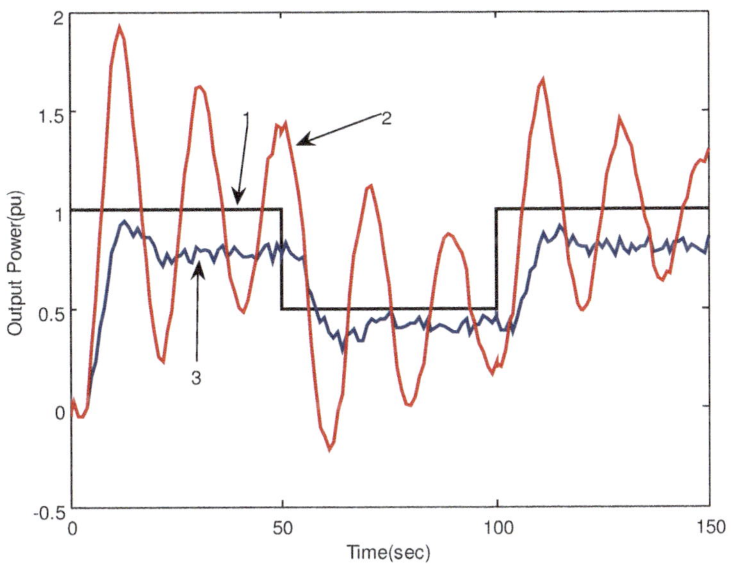

Fig. 14 PID Controller Responses to Set-Point Control with output noise. 1-Setpoint Reference 2- Plant Output Response 3- NN Output Response

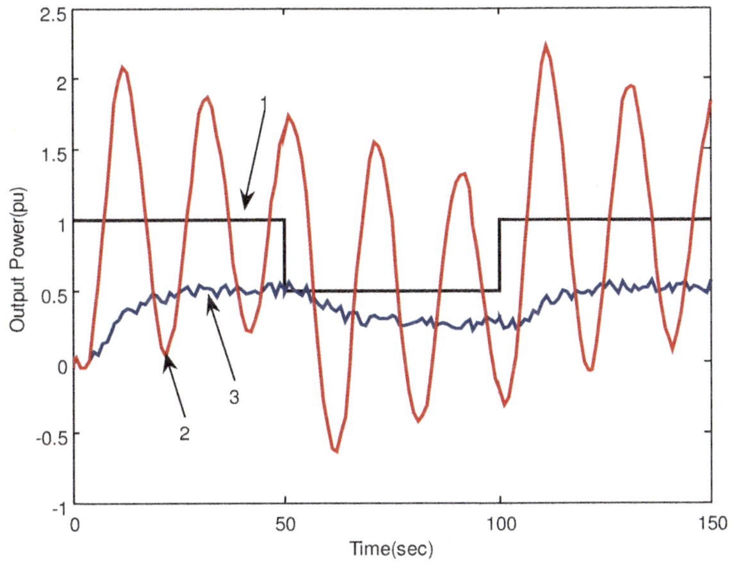

Fig. 15 Neuro-Controller with BPP Learning Responses with output noise. 1-Setpoint Reference 2- Plant Output Response 3- NN Output Response

6 Conclusions

This chapter aims to highlight the self-tuning control applications of an efficient neural network structure derived from *wavenets* —a wavelet theory. A relevant network topology without primarily using trial-and-error methods can be demonstrated by *Wavenet* algorithms. *Wavenet*-based controllers are capable of improving the performance of the trained network on behalf of the quick convergence, least amount of changes between runs, robustness to noise interference, and high complex ability to learn and track of unknown or undefined complex systems. Three control schemes are illustrated: a plant model and a neuro-identifier, applied to construct adaptive controllers, a PID controller derived from the self-tuning *wavenet* adaptations, and a traditional neuro-control scheme using the feed-forward neural network structure along with back-propagation (BPP) algorithms. The latter is used to make a base line comparison with *wavenets.* The first two schemes have different advantages. The PID controller is simple to configure since it is needless to a process model. It is easy to adapt *wavenet* algorithms to the existing PID controllers and plants because of their ability in providing simple transition changes to plant modifications. The first controller involves less parameter than the PID controller, which requires extra PID parameters. The results of simulation studies are very similar in both of the schemes, but the PID scheme has a tendency to be more robust and less sensitive. The first scheme tends to provide faster tracking adjustment to control changes. Finally, it is demonstrated that the conventional scheme with BPP requires a longer time adapting to changes and has a poor performance to systems with added noise. Research studies show the worst situation to all of the control schemes, in terms of MSE, arises when noise is contaminating the output port of the plant systems.

Acknowledgments

The authors were supported in part by a Research grant From Shahid Beheshti University, G.C. The authors are indebted to the editor and referees for greatly improving this chapter.

References

1. Li, X., Cao, G., Zhu, X.: Modeling and control of PEMFC based on least squares support vector machines. Energy Conversion and Management 47, 1032–1050 (2006)
2. Zhang, L., Pan, M., Quan, S.: Model Predictive Control of Water Management in PEMFC. Journal of Power Sources, doi:10.1016/j.jpowsour.2008.01.088 (accepted paper)
3. Nguyen, T., Knobbe, M.W.: A liquid water management strategy for PEM fuel cell stacks. Journal of Power Sources 114(1), 70–79 (2003)

4. Chen, D.M., Peng, H.: Proc. of the 2004 American Control conference, pp. 822–827 (2004)
5. Abtahi, H., Zilouchian, A., Saengrung, A.: Water management of PEM fuel cells using fuzzy logic controller system. In: Proc. of the IEEE International Conference on Systems, Man and Cybernetics, vol. 4, pp. 3486–3490 (2005)
6. Guo, L.: Predict Control algorithm and application of output power of fuel cells. Chinese Journal of Power Source 28(6) (2004)
7. Ren, Y., Guang-Yi, C., Xin-Jian, Z.: Predictive Control of Proton Exchange Membrane Fuel Cell (PEMFC) Based on Support Vector Regression Machine. In: Proc. of the Fourth International Conference on Machine Learning and Cybernetics (2005)
8. Sridhar, P., Perumal, R., Rajalakshmi, N., Raja, M., Dhathathreyan, K.S.: Humidification studies on polymer electrolyte membrane fuel cell. Journal of Power Sources 101(1), 72–78 (2001)
9. Narendra, K.S., Parthasarathy, K.: Identification and control of dynmical systems using neural networks. IEEE Trans. on Neural Networks 1(1), 4–27 (1990)
10. Dong, W., Cao, G.Y., Zhu, X.-J.: Nonlinear Modeling and Control based on Adaptive Fuzzy Technique for PEMFC. Control and Intelligent Systems Journal (2004), http://www.actapress.com/Content_Of_Journal.aspx?JournalID=64(201)
11. Zhang, L., Pan, M., Quan, S., Chen, Q., Shi, Y.: Adaptive Neural Control Based on PEMFC Hybrid Modeling. In: Proc. of the 6th World Congress on Intelligent Control and Automation, pp. 8319–8323 (2006)
12. Sedighizadeh, M., Arzaghi-Harris, D.: A Neuro Adaptive Control Strategy for Movable Power Source of Proton Exchange Membrane Fuel Cell Using Wavelets. In: Proc. of the 41st International Universities Power Engineering Conference (UPEC), vol. 2, pp. 545–549 (2006)
13. Li, Y., Wang, H.: Using Artificial Neural Network to Control the Temperature of Fuel Cell. In: Proc. of the International Conference on Communications, Circuits and Systems Proceedings, vol. 3, pp. 2159–2162 (2006)
14. Lekutai, G., VanLandingham, H.F.: Self-tuning control of nonlinear systems using neural network adaptive frame wavelets. In: Proc. of the IEEE International Conference on Systems, Man, and Cybernetics, vol. 2, pp. 1017–1022 (1997)
15. Wang, J., Wang, F., Zhang, J., Zhang, J.: Intelligent Controller Using Neural Network. Intelligent Manufacturing. In: Yang, S.-Z., Zhou, J., Li, C.-G. (eds.) Proc. of SPIE, vol. 2620, pp. 755–761 (1995)

Erratum to: Network Complexity Analysis of Multilayer Feedforward Artifical Neural Networks

Helen Yu

J. Schumann, Y. Liu (Eds.): Appl. of Neural Networks in High Assur. Sys., SCI 268, pp. 41–55.
springerlink.com

DOI 10.1007/978-3-642-10690-3_12

In the original version, the author name "Helen Yu" appearing in the Table of Contents and in the chapter is incorrect. The correct name is "Xiao-Hua Yu".

The original online version for this chapter can be found at
http://dx.doi.org/10.1007/978-3-642-10690-3_3

GPSR Compliance

The European Union's (EU) General Product Safety Regulation (GPSR) is a set of rules that requires consumer products to be safe and our obligations to ensure this.

If you have any concerns about our products, you can contact us on ProductSafety@springernature.com

In case Publisher is established outside the EU, the EU authorized representative is:

Springer Nature Customer Service Center GmbH
Europaplatz 3
69115 Heidelberg, Germany

Batch number: 10755466

Printed by Printforce, the Netherlands